Urban Landscape Entomology

Urban Landscape Ecology

Urban Landscape Entomology

David W. Held, BS, MS, PhD
Professor
Department of Entomology and Plant Pathology
Auburn University
Auburn, AL
United States

ELSEVIER

ACADEMIC PRESS
An imprint of Elsevier

Academic Press is an imprint of Elsevier
125 London Wall, London EC2Y 5AS, United Kingdom
525 B Street, Suite 1650, San Diego, CA 92101, United States
50 Hampshire Street, 5th Floor, Cambridge, MA 02139, United States
The Boulevard, Langford Lane, Kidlington, Oxford OX5 1GB, United Kingdom

Notices
Knowledge and best practice in this field are constantly changing. As new research and experience broaden our understanding, changes in research methods, professional practices, or medical treatment may become necessary.

Practitioners and researchers must always rely on their own experience and knowledge in evaluating and using any information, methods, compounds, or experiments described herein. In using such information or methods they should be mindful of their own safety and the safety of others, including parties for whom they have a professional responsibility.

To the fullest extent of the law, neither the Publisher nor the authors, contributors, or editors, assume any liability for any injury and/or damage to persons or property as a matter of products liability, negligence or otherwise, or from any use or operation of any methods, products, instructions, or ideas contained in the material herein.

Library of Congress Cataloging-in-Publication Data
A catalog record for this book is available from the Library of Congress

British Library Cataloguing-in-Publication Data
A catalogue record for this book is available from the British Library

ISBN: 978-0-12-813071-1

For information on all Academic Press publications visit our website at
https://www.elsevier.com/books-and-journals

Publisher: Charlotte Cockle
Acquisition Editor: Anna Valutkevich
Editorial Project Manager: Pat Gonzalez
Production Project Manager: Kiruthika Govindaraju
Cover Designer: Alan Studholme

Typeset by TNQ Technologies

Contents

Preface

This project originated in the classrooms and meeting halls where I have spoken to numerous students and professionals over the years. Students come to campus to be trained in the principles and hypotheses needed for their careers in ornamental horticulture and turfgrass management. In the courses I teach, this is handled by reading research papers on these topics and hypotheses. However, during my time in Extension, I observed that few professionals had neither access to nor time to read the primary research journals where these ideas are deposited and debated. As a result, I noticed the bottom rungs either were missing or have slowly disappeared from this ladder linking entomological and ecological principles to practice. Those bottom rungs are the basic concepts often cited in research papers but not commonly explained in practical ways for students or Extension clientele. About 7 years ago, I included a few of these principles in a presentation and explained them in ways I thought professional landscape or turfgrass managers could understand. I expected mixed reviews considering most professional audiences attend looking for shovel- and sprayer-ready ideas, and not hypotheses. But, the responses to these presentations were great. Following that first presentation, a few attendees asked thoughtful questions and provided real-world examples that either supported or refuted what was presented. As a researcher and educator, I was pumped up and encouraged by this response. Coincident with this, I noticed that the book format was not dead, despite the ubiquity of good information. Several colleagues released books targeting professionals and students that were well received and successful. I wondered if a book covering the breadth of topics in urban landscape entomology in a digestible format would equally resonate with these audiences. Therefore, this book was proposed and written to put bottom rungs on the ladder for the professional arborist, landscape manager, turfgrass manager, Extension agent, or horticulture student.

The Table of Contents is based on the syllabus from my *Landscape Entomology* course. For the course and this book, I picked topics that I perceived were murky, misunderstood, or controversial and asked myself how might I review and teach those ideas to foster understanding. In a laboratory session with my course, I can teach insect identification, but my goal with the book was not to answer every, "What's this bug?" question. I felt the other books and book chapters preceding this one already addressed insect diagnostics. Only three of the nine chapters are devoted to diagnostics, sampling, and pest biology. The remaining six chapters present topics in urban landscape entomology such as urbanization effects on plants and insects; insects and mites and their interactions with plants as influenced by fertilization and stress; and integrated pest management principles. Each chapter includes three to four bulleted "Highlights," or key points discussed in that chapter. These were intended to provide a quick reference to the main ideas for the chapter. Each chapter is also fully referenced with a current literature review for readers wanting further study on a particular topic. I know you will find this book a useful, practical introduction to the principles and practices in urban landscape entomology.

David W. Held
May 2019

Acknowledgments

Several times while working on this book, I questioned my sanity and ability to finish such a large project. At those times I relied on the support and encouragement of family and friends. My wife Michelle, and children Mary and Michael, thankfully allowed me to be a ghost at home in the final 3 months of preparing this book. At times they were also employed to proofread sections of chapters or to check references. Furthermore, my family followed my career through three states and granted me leniency to travel out of state and country for teaching and career development. Some valuable experiences shared in this book were gained on those trips. Thanks also to my daughter for accepting the challenge to develop two images in Chapter 1 as part of her graphic design course at Auburn High School. Mr. Clay Cox, her graphics design instructor, provided helpful input during that process. Many thanks are due to friends who provided shots of encouragement while I was writing. I leaned on those words late at night to keep me going.

Dr. Daniel Potter started me on the path that led to this point. His teaching, mentorship, and investment in me began in 1995 when I took a research position in his lab. Dr. Potter continues to provide me with advice and encouragement whenever I call. Since this is single-authored volume, I relied on feedback from reviewers to check the presentation and content. Dr. Daniel Herms (Vice President of Research and Development, The Davey Tree Expert Company, Kent, OH), Kendra Carson, Elijah Carroll, Evan Kilburn, and Oluwatomi Daniel Ibiyemi read earlier drafts of chapters and provided helpful feedback.

This project would not be possible without the nearly 700 citations of scholarly books and papers spanning >70 years. Among them are Entomology Masters or Doctoral degree students who passed through my lab. There is no clear point of origin for the field of landscape entomology in the United States. However, papers by Dr. Ephraim Porter Felt in the early 1900s provided early observations and life history information on many common pests of woody plants in the United States. A few of Dr. Felt's books are on my bookshelf, and I cite his work on galls in Chapter 7. Thanks to all of the cited researchers for your contributions to this field.

Pat Gonzalez and the team at Elsevier/Academic Press provided invaluable support for this project. The cover art and Fig. 7.1 are excellent additions to this book. Thanks again for helping this project to be completed with excellence and on time. I am sure that every new book project brings the same reoccurring headaches and questions from new authors or editors. Your patience through the process was much appreciated. Pokonobe Associates, owner of the JENGA Brand, granted permission to use the brand name and image in the book. That game is a powerful visualization of key concepts in Chapter 1.

Chapter 1

Introduction

Urbanization: trends and relevance

Urbanization creates effects on the land, plants, and pests in ways unlike other plant systems (agriculture or forestry). Urbanization is the shifting of people into or nearby cities, but often for better social or economic opportunities. The United States generally lags behind most of Asia and Africa in present and future urbanization (1). This lag is likely due to our relatively new colonization compared to Europe and Asia, but several United States megacities (e.g., New York, Los Angeles) are rapidly catching up with international megacities. Many sources have estimated the rapid pace of urbanization in the United States with the UN data being the most cited. In the 40-year period from 2010 to 2050, urban populations are predicted to increase 10%. The United States is already 82% urban and, by 2050, should grow to 90% of the population in urban\suburban areas. The same trend is occurring globally. Megacities are growing and 68% of the global population is expected to be urban within the next 30 years (2).

If you have witnessed urbanization firsthand in the United States, it was likely agricultural or tree crop (forestry or tree fruit) production lands converted to various urban developments. I call this the *crops to cul-de-sacs effect*. Losses of land to urban development are considered permanent as it is uncommon but not impossible (3) to convert urban spaces back to natural or agricultural lands. The lower 48 states have a finite land mass of 768.9 million ha (1.9 billion acres). From 1982 to 2012, urban development in the United States increased by 17 million ha (42.2 million acres) (Fig. 1.1). In 2012, urban areas accounted for 5.86% (46,174,631 ha or 114.1 million acres) of the entire United States land cover (4). These statistics are also cited as the basis for claims that urbanization is a threat to agricultural productivity and global biodiversity (5). As we replace arable lands with houses, golf courses, or shopping malls, a greater burden of food and fiber productivity is shifted to less land. The average farmer in 2012 was feeding about 100 people in the United States (3.2 million farmers feeding a population of >300 million people) (6,7). The ratio of people in the United States to farmers increases with every Census of Agriculture and is driving agricultural intensification, or more productivity from less land.

Urban Landscape Entomology. https://doi.org/10.1016/B978-0-12-813071-1.00001-4

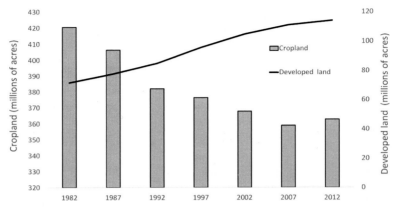

FIGURE 1.1 The continental United States has 1.9 billion acres of land cover. Of this, an increasing portion is becoming urban mainly at the expense of cropland (*4*).

BOX 1.1 Plant communities shape insect communities.

Plants are the primary food source in these communities. As we will see later, these effects are well documented for plant-feeding insects as well as insect natural enemies and pollinators. These effects in research papers are called "bottom-up" effects or forces (for more information, see Ref. *(8)*).

Landscape structure, both living and hardscape components, are drivers of ecological change in urban landscapes (Box 1.1). As plant species and abundance change, insect populations change. The addition of hardscape (buildings) or impervious surfaces (roads, sidewalks) covers soil or removes topsoil, changes temperatures, and the biological components of landscapes are impacted by these changes (*9*). Understanding these interactions and their influences on plants, insects (pests and beneficials), and people (socioeconomics) are common themes in the broader field of landscape ecology.

Have you heard someone suggest that they see fewer fireflies (*9*), dragonflies, or other insects today relative to their childhood or even 10 years ago? Or maybe you have read a story on the windshield effect (*10*), where people perceive fewer insects by the number of splats on their windshield? Some popular press writers have even dubbed these documented losses of insect biodiversity as an "insect apocalypse" (*11*). Habitat destruction through urbanization is often implicated in these effects on insects (*5*). To document these effects, scientists commonly compare urban to nonurban locations to discern the effects of urbanization. If there are significant differences between the two locations, then an urbanization effect is affirmed and the causes can be further investigated. A few studies, however, have carefully documented changes in insect communities at one location over time. Using historic records or insect collections in museums, it is possible to estimate local extinctions of a group or species of insects over time. Annual data collection is the basis for two well-publicized North American projects, Monarchwatch.org and lostladybug.org. Europe, however, has more years of urban development and data collection. Rome, one of the oldest cities in the Western world, is an excellent case study. As Rome urbanized, species of darkling beetles (Tenebrionidae) became less dense and isolated and some became locally extinct. Darkling beetles are not pests of landscape plants, but a common family that provides biological services like decomposition, scavenging, and biological control (*12,13*).

Urban landscapes: interactions of people, plants, and insects

How can urbanization cause such effects? Let us begin by imagining a fictitious expanse of land, fully or partially forested, located somewhere in eastern North America. If we go back in time before European colonization, plants would be exclusively native trees, shrubs, and groundcovers (grasses or other herbaceous plants) adapted for this location, with biological associations among insects, birds, and mammals. Large trees like mature oaks, maples, beech, or evergreens may represent the overstory, which is the highest canopy layer in a forest. The understory, the tree layer below the overstory, includes the younger saplings of overstory trees, and understory trees like dogwoods or redbuds. The shrub layer could contain oakleaf hydrangeas, bayberry, rhododendrons, or hollies. The ground layer would be a mixture of herbaceous perennials, annuals, vines, and decomposing leaves.

Now, overlay onto this plant background some ecological processes and relationships (Fig. 1.2). The plants represent the food-making component called primary producers. Plants harvest the sun's energy and carbon dioxide to generate biomass (leaves, stems, flowers, fruit, roots). Primary consumers get their energy, and the ability to grow and develop, from the primary producers. The insects tend to dominate the primary consumers consuming mainly leaves but also other plant parts like pollen. A third level called decomposers would recycle the unused plant biomass and animal waste. Among insects, plant feeders (herbivores) can either be specialists, using a few plants as hosts, or generalists, using a wide range of hosts. Azalea lace bugs (*Stephanitis pyrioides*) are examples of specialists, and Japanese beetles (*Popillia japonica*) are examples of generalist herbivores. The next level, called the secondary consumers, are insects and other animals that eat the primary consumers. They are generically called natural enemies. The secondary consumers get their energy and biomass from consuming primary consumers or by consuming individuals at their same level. This group of animals works to balance the populations of plant feeders. As with herbivores, insect natural

FIGURE 1.2   A triangle is commonly used to represent trophic levels. In urban landscapes, plants are the primary producers represented as the large base. Primary consumers would use plant parts (leaves, roots, pollen, nectar) for food. Secondary consumers at the top represents the natural enemies. Each block in the triangle is dependent on the block below and provides support for the block above. *Figure credit: M. Held.*

enemies can also be further classified into specialists and generalists depending on the number of insects they use for food. Each level of the tower or pyramid builds upon the level below it. These levels, called trophic levels, indicate the level at which an animal feeds. Now our hypothetical tract of land contains plant and animal communities reproducing, developing, and inter-acting with each other.

Each of the levels is present and represented by a one or more species, further represented by multiple individuals of the same species. We can measure and characterize the species richness (number of species present) and evenness (how similar are the number of individuals across species) of plants and insects in this tract of land. Species in our pretend community are con-tained within populations, and year to year the numbers could increase or decrease. In time, populations would reach what is called the equilibrium position where the yearly ebb and flow of numbers produce a long-term average. Producers and consumers represent trophic levels in a biological system with many points of connection. In the case of insect herbivores, the plants used by these consumers would represent a check point or limit on the populations of herbivores. This should be intuitive as there are only so many plants, or plant parts that can support plant-feeding insects. This ability of a habitat or environment to support populations is the carrying capacity. It represents the maximum population that can be supported by that particular environment. There are other points of connections between producers, consumers, and secondary consumers. For example, aphids present on the leaves of the oak trees are eaten by predatory lady beetles and lace wings, and larvae of the parasitoid *Aphelinus*. The adult forms of these beneficial insects can supplement their diet with pollen or nectar from the flowers of the shrubs or groundcover layers. These connections go largely unnoticed by humans from year to year but are key to the biological function and stability of this community.

The connections in our pretend community are not unlike bricks in a wall or blocks in the game Jenga® (*14*). Jenga® uses small wood blocks which fit together tightly into a tower (Fig. 1.3). Each block is square on the tops, bottoms, and sides to provide the most support for the adjacent blocks. The block tower represents our community and each block is a species in our community. Now, let us say that we develop this tract of land for housing, shopping, or sports fields. Most of the understory, shrub, and groundcover layers are removed, but we retain most of the large trees. *What does this do to our community?* Going back to Jenga®, the tower can typically stand with some of the blocks removed, that is the purpose of the game. Just as there may be holes in the tower, there are holes in our community created by urbanization. Houses, roads, lawns, and new plants get installed. However, most of

FIGURE 1.3 The block game Jenga® provides a visualization of the relationships between plants, beneficial insects, and plant-feeding insects in urban landscapes (*14*). Connections in an established forest or wooded lot may look like the tower on the left. The tower on the right shows missing and irregularly shaped blocks. These spaces and odd-shaped blocks are where endemic plants or insects may be removed or replaced by nonnatives. The tower still may be able to stand but the stability is impacted. Jenga® is a trademark of Pokonobe Associates. Used with permission. All rights reserved. ®refers to U.S.A. trademarks. *Figure credit: M. Held*

the available plant choices are nonnative (*15*) but perhaps related to native counterparts. This means that we may replant using a nonnative kousa dogwood or azalea, rather than the native Florida dogwood or azalea. The nonnative dogwood and azaleas are like adding pieces back to the block tower; however, those pieces may not have the shape and length of the original pieces. Some may be shorter, some not as thick. They are still useful to support the community (tower) but are not exactly the same as the original species (*16,17*). The groundcover layer now becomes mostly mown turfgrass. Turfgrass is now the largest single plant component by area in the United States (*18*). The flowering groundcover of native plants now changes to the annuals and perennials available at the local garden center or home store. Some, but not all of them, may still serve the original function of providing pollen and nectar to beneficial insects. More blocks are added back into our tower, but again they may lack the shape or size of the original blocks. As this landscape ages, the connections in our community likely change, but it may take time to create the multiple points of connection from the original landscape. Some ecologists argue that the original complexity of connections is forever lost as a result of urbanization.

A real and starker example would be if the entire tract of land was cleared and the new community was started from a "blank slate." In this case, a contractor sometimes guided by a landscape designer would reinvent a landscape based on human utility. The plants, now installed in patches that lack resemblance to natural densities or species richness, function for mainly human utility as screens, hedges, or to direct people or traffic, with limited or different ecological functions. For example, human utility likes to plant overstory and understory trees in rows at the edge of a lot or property, a configuration that never existed in this tract prior to human intervention. Newly installed trees, oaks and maples, destined to be the new overstory, will remain understory trees for many years during which time that important layer will be absent. In this case, the tower is being rebuilt but now with different blocks, more variable than the original. This degree of change is likely driving some of my colleagues to suggest urban landscapes or green spaces are "disasters by design" (*19*). Although largely pessimistic, the optimistic could argue to change the design to make improvements. Fortunately, there are books (*20*) and plenty of articles in landscape planning journals (e.g., Ref. (*21*)) that contain design concepts and best practices to preserve biodiversity or conserve beneficial insects in urban green spaces. Principles of ecological landscape design are not a primary focus of this book. But, be encouraged that landscape architects appear more aware now than before of putting these principles into practice for more sustainable urban landscapes. The remainder of this book will build upon the examples and concepts outlined in this chapter and focus on what is presently known, and the gaps in our understanding of urban landscape entomology.

Highlights

- Urbanization in the United States is increasing largely at the expense of farmland.
- Urbanization is considered an important threat to insect biodiversity.
- Urbanization changes the plant community and the ecological relationships between plants and insects.

References

1. Bren d'Amour, C.; Reitsma, F.; Baiocchi, G.; Barthel, S.; Güneralp, B.; Erb, K. H.; Haberl, H.; Creutzig, F.; Seto, K. C. Future Urban Land Expansion and Implications for Global Croplands. *Proceeding of the National Academy of Science USA* **2017,** *114* (34), 8939–8944.
2. United Nations, Department of Economic and Social Affairs/Population Division, World Urbanization Prospects: The 2018 Revision https://population.un.org/wup/Publications/.
3. Gardiner, M. M.; Burkman, C. E.; Prajzner, S. P. The Value of Urban Vacant Land to Support Arthropod Biodiversity and Ecosystem Services. *Environmental Entomology* **2013,** *42* (6), 1123–1136.
4. United States Department of Agriculture and Natural Resources Conservation Service. *National Resources Inventory Summary Report,* 2012. https://www.nrcs.usda.gov/Internet/FSE_DOCUMENTS/nrcseprd396218.pdf/.
5. Sánchez-Bayo, F.; Wyckhuys, K. A. G. Worldwide Decline of the Entomofauna: A Review of its Drivers. *Biological Conservation* **2019,** *232,* 8–27.
6. United States Department of Agriculture 2012 Census of Agriculture, https://www.agcensus.usda.gov/Publications/2012/Online_Resources/Highlights/Farm_Demographics/.
7. United States Census Bureau, American FactFinder, 2010 Census data, https://factfinder.census.gov/faces/nav/jsf/pages/index.xhtml.
8. Vidal, M. C.; Murphy, S. M. Bottom-up vs. Top-Down Effects on Terrestrial Insect Herbivores: a Meta-Analysis. *Ecology Letters* **2018,** *21* (1), 138–150.
9. Raupp, M. J.; Shrewsbury, P. M.; Herms, D. A. Ecology of Herbivorous Arthropods in Urban Landscapes. *Annual Review of Entomology* **2010,** *55,* 19–38.
10. Vogel, G. Where Have All the Insects Gone? *Science* **2017,** *356* (6338), 576–579.
11. Jarvis, B. The Insect Apocalypse Is Here. What Does it Mean for the Rest of Life on Earth? *The New York Times* **November 27, 2018.**
12. Fattorini, S. Insect Extinction by Urbanization: A Long Term Study in Rome. *Biological Conservation* **2011,** *144* (1), 370–375.
13. Fattorini, S. Faunistic Knowledge and Insect Species Loss in an Urban Area: The Tenebrionid Beetles of Rome. *Journal of Insect Conservation* **2013,** *17,* 637–643.
14. de Ruiter, P. C.; Wolters, V.; Moore, J. C.; Winemiller, K. O. Food Web Ecology: Playing Jenga and Beyond. *Science* **2005,** *309,* 68–71.
15. Coombs, G.; Gilchrist, D., Native and Invasive Plants Sold by the Mid-Atlantic Nursery Industry: A Baseline for Future Comparisons. Available from: https://mtcubacenter.org/action/%20nurserysurvey.
16. Tallamy, D. W.; Shropshire, K. J. Ranking Lepidopteran Use of Native versus Introduced Plants. *Conservation Biology* **2009,** *23* (4), 941–947.

17. Clem, S. C.; Held, D. W. Species Richness of Eruciform Larvae Associated with Native and Alien Plants in the Southeastern United States. *Journal of Insect Conservation* **2015,** *19,* 987–997.

18. Milesi, C.; Running, S. W.; Elvidge, C. D.; Dietz, J. B.; Tuttle, B. T.; Nemani, R. R. Mapping and Modeling the Biogeochemical Cycling of Turf Grasses in the United States. *Environmental Management* **2005,** *36* (3), 426–438.

19. Raupp, M. J.; Shrewsbury, P. M.; Herms, D. A. Disasters by Design: Outbreaks along Urban Gradients. In *Insect Outbreaks Revisited;* Barbosa, P., Letourneau, D. K., Agrawal, A. A., Eds.; Wiley-Blackwell: UK, 2012; pp 313–340.

20. Muller, N., Werner, P., Kelcey, J. G., Eds. *Urban Biodiversity and Design;* Wiley-Blackwell: UK, 2010.

21. Garrard, G. E.; Williams, N. S. G.; Mata, L.; Thomas, J.; Bekessy, S. A. Biodiversity Sensitive Urban Design. *Conservation Letters* **2018,** *11* (2), 1–10.

Chapter 2

Landscape structure and complexity

In Chapter 1, we laid a foundation on which we can now build the components found in urban landscapes. The new landscape we created in Chapter 1 essentially is a patch within a matrix. A matrix is the larger, more extensive landscape component in which a small-scale urban landscape occurs. If you are at the center of a large city, your matrix consists of man-made structures and roads. In rural areas, the matrix can be the native forest or perhaps agricultural production land. Within the matrix, there are patches and corridors. Large patches could be subdivisions, golf courses, parks, or retail developments. Small patches may be plantings of azaleas or roses in a residential landscape. Patches can be made by planting something new or by isolating an area through development. If the development of the new landscape from Chapter 1 left a grouping of native trees in the middle of that site it would be a remnant patch. That remnant patch was once connected but is now isolated and disconnected ecologically. Patches in the urban landscape are easy to visualize: think of the common subdivision used throughout the United States. The process, called fragmentation, creates patches (Fig. 2.1). Some patches maintain connectivity through passages called corridors. Corridors are like roads through which wildlife can move. In highly urbanized areas, these corridors also serve as one of the few places in these landscapes where the seeds of wild plants can germinate and populations expand independent of human activity. Several neighborhoods in Auburn, Alabama, have remnant patches of endemic trees throughout that provide corridors for wildlife. Those corridors provide passage for deer, fox, and other vertebrate animals. These patches can be considered biological islands following the principles of colonization and species richness developed from research on the colonization of real islands (i.e., island biogeography) (1). Intuitively, island size should influence species with larger biological islands having greater species richness (species to area relationship). *But, do urban islands in landscapes follow principles determined for real islands?* One review (2) noted that 68% of the case studies in urban green spaces show an increase in insect species richness with an increasing or larger patch size. This also means the opposite, that smaller patches in urban landscapes should have lower species richness and

Urban Landscape Entomology. https://doi.org/10.1016/B978-0-12-813071-1.00002-6

Extent of Development

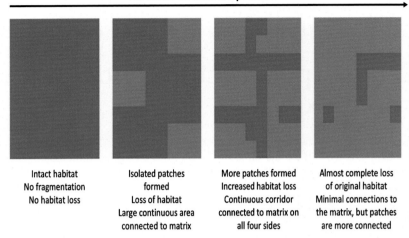

| Intact habitat
No fragmentation
No habitat loss | Isolated patches
formed
Loss of habitat
Large continuous area
connected to matrix | More patches formed
Increased habitat loss
Continuous corridor
connected to matrix on
all four sides | Almost complete loss
of original habitat
Minimal connections to
the matrix, but patches
are more connected |

FIGURE 2.1   Each urban landscape exists within a matrix (green). As development occurs, patches are created within the matrix. Depending on the extent of development there may be a continuous corridor. However, development may occur until the original matrix is mostly replaced by interconnected patches that have little physical or ecological connection to the original landscape or the matrix.

abundance, is also true. As patches get smaller, however, it is not like getting a smaller piece of the larger pie that has the same filling of the entire pie. Pie pieces, like biological islands, may be cut in a way to exclude certain parts of the filling. Likewise, smaller biological islands of varying sizes will exclude certain species that just cannot survive once the patch gets below a particular size threshold (*3*). Research on real islands also predicts that islands farther from new colonists (insects) should not be as species rich as those closer to other islands or a mainland (*1*). This prediction is also supported by case studies in urban green spaces (*2*).

As patches are created, edges or margins are by-products of changes in habitats. Edges occur commonly in urban landscapes. A few examples include the conspicuous edge where a forested patch meets the grass in a residential landscape but it can also be where the tall grass of the rough meets the fairway height grass on a golf course. Annual bluegrass weevils, a grass feeding insect, have a behavioral response to the edge created by different mowing heights on a golf course. The adult weevils accumulate along these edges in spring which produces greater numbers of larvae in the grass in those areas (*4*). Edges can also cause unique effects on animal populations, including insects. One such effect, called spillover, is where insect herbivores, natural enemies, or pollinators move between habitats. In the case of herbivores, it may be an herbivore moving from a host plant in the matrix to a related host plant now present in the landscape (patch). Likewise, pests from urban landscapes can spillover into natural habitats. One devastating example is the spillover of the hemlock woolly adelgid (*Adelges tsugae*). Adelgids are aphidlike, small insects that suck sap from evergreen trees: in this case hemlocks. Woolly adelgids on hemlocks (*Tsuga* spp.) planted in a private landscape in Virginia spilled over into the native hemlock stands in the eastern United States (5). The spillover effects of this one insect have impacted hemlock survival, forest nitrogen cycles, and populations of birds, spiders, and salamanders in 17 states (5).

Fragmentation and spatial scales

Urban and suburban areas, although highly fragmented, have distinct structure and functions. Structure is evident in visible elements (components) and their arrangement (configuration). A common pastime I have while flying is looking out the airplane window at low altitudes and seeing a "bird's-eye" view of landscapes. The next time you fly, awkwardly lean in front of the person in the window seat and look out. The structure of the landscape is amazingly revealed from the air. There are hardscape features like sidewalks, streets, homes, as well as the biological components of tree and shrubs. Depending on the altitude, you can clearly see that there are obvious spatial scales to the landscape. There is the city level spatial scale usually measured in hundreds of square kilometers. The town of Auburn, Alabama, for example, is about $150 \, km^2$ (58 sq. mile). Within a city, there are districts or neighborhoods having a spatial scale range from 1 to $10 \, km^2$ (39 sq. miles). As we zoom in closer, there are features like parks, streets, or golf courses on sites about $0.5-1 \, km^2$, but not quite as small as the average subdivision lot which is $0.25-1$ acre ($0.001-0.004 \, km^2$). This discussion of geography is important for urban insects which tend to be managed at the smallest and most localized spatial scales. For example, if lace bugs outbreak on azaleas, fall armyworms or chinch bugs in a lawn, a homeowner or landscape manager treats those populations as unique, isolated from others on the same street or in the same neighborhood. The scale on which management is applied is not always the ecologically important spatial scale for urban insects. To understand the ecology of urban landscapes, you must recognize that, although humans are bound by property lines or municipal boundaries, insects are not. Therefore, not all insects that you notice in a small urban space originated from that space nor do they necessarily have to be residents of that space.

Spatial scales and insect movement

Landscape ecologists have excellent data on spatial relationships of large animals relative to urbanization. However, principles in landscape insects and wildlife ecology are not always directly applicable to urban landscapes because insects and plants require smaller areas than larger animals to support viable populations. Insects can develop in a smaller area but that does not mean they are limited to the smallest spatial scales for their entire life. Let us consider the larvae of Monarch (*Danaus plexippus*) caterpillars. A small planting of milkweed (*Asclepias* spp.) could support the larva but, for persistence of the species, the adults migrate to Mexico annually and return to the United States the following spring. This is just one of several insects that move across larger spatial scales usually in the adult life stage. Globally, migratory locusts often account for the most damage of any migratory pest. Large, adult insects with wings can move long distances or in and out of a given habitat or patch. Long-range movements of urban insects are not common, but here are a few examples of this phenomenon among adult insects common in urban landscapes.

- Honey bees (*Apis mellifera*) can forage for 1.6–4.8 km (1–3 miles) from a colony in 1 day (*6*).
- Japanese beetle adults can disperse as much as 8 km (5 miles) a year from the point where they developed as grubs (*7*).
- Moths of the black cutworm are migratory and disperse southward from the northern states in the fall. The pupae of this species only survive the winter in southern states. In spring, moths move northward on southerly airflow (*8*).
- Similar to black cutworms, moths of the fall armyworm (*Spodoptera frugiperda*) (*9*) and adult potato leafhoppers (*Empoasca fabae*) migrate to northern states each spring from Florida and the other Gulf states (*10*).
- Emerging scale insects are a few millimeters (0.125–0.5 in.) in length, but can be dispersed from infested trees over long distances by wind, other insects, or animals. Crawlers can quickly (1–5 min) "board" larger insects like flies or beetles and remain attached for up to 1 h of flight after boarding the insect carrier (*11*). Using mild wind currents (<1 m per sec [~2 miles per hour]), crawlers can disperse locally (33 ft) (*12*) or regionally up to 3–6 km (2–3 miles) from infested trees (*13*).

During my time as an Extension specialist, homeowners would commonly suggest that insect damage or insects appeared overnight. Knowing that it could not be "spontaneous generation," local movement of caterpillars or adult insects was the logical explanation. Movement within or between urban landscapes is common and can be one explanation for an apparent "instant infestation." Here are a few examples of local movements of adult and immature stages of insects found in urban landscapes.

- In the fall, adults of the annual bluegrass weevil and the black turfgrass ataenius move from the golf course into adjacent wooded areas (*14,15*). Adult annual bluegrass weevils can be up to 60 m (197 ft) from nearby golf course fairways (*15*). In spring, the same adults move back to the short-mown grass on fairways to feed and lay eggs.
- Adult southern chinch bugs can fly but dispersal is common on the ground. These relatively small insects can move (>30 m [>100 ft]) on the ground between lawns in urban landscapes in a matter of hours (*16*).

- Caterpillars can disperse locally usually as newly hatched larvae. However, once eastern tent caterpillars defoliate a tree, they can disperse, using silk or crawling, to nearby host plants in the landscape. On the ground, larvae can crawl 10−15 m (≥30 ft) to find a new host or a site to pupate (*17*).
- Bagworm larvae also disperse using silk. The dispersal distances of larvae on silk depend on wind speed but they can be captured up to 30−75 m (98−245 ft) from an infested host tree (*18−20*).
- On golf courses, black cutworm larvae can crawl up to 21 m (70 ft) in one night when moving from the taller grass in the rough to greens and tees (*21*).
- Imported fire ant queens, in their winged or alate form, can disperse 2 km (1.2 miles) from the nest during nuptial flights (*22*).
- In short-mown bermudagrass, fall armyworm larvae can disperse 151 cm (5 ft) in a few hours, easily capable of reaching a neighboring lawn within a day (*Carroll and Held, unpublished data*).

Influence of local landscape components on insects

At local spatial scales (under 0.5−1 km^2), landscape components can in-fluence the distribution of insect populations. Distribution, where insects are located, is different from density, which tells you how many are present. Insects in an urban landscape are more commonly in clumps, or areas where individuals are locally dense. For many landscape pests, the local configuration or occurrence of plants or abiotic (nonliving) components will influence pest distributions (where you can find them). Distributions are useful for sampling and detection (see Chapter 5). It is easier to manage insects if you know where, or under what conditions, you are most likely to them.

We will first consider the concept of structural complexity, which is a measurement of the occurrence of vegetation in multiple layers or strata (Fig. 2.2). In Chapter 1, we defined the layers or strata—overstory, un-derstory, shrub layer, annual and perennial layer, and groundcover possible in urban landscapes (*23*). As you can imagine, urban landscapes may have all or just some of these layers. Landscapes with more layers or more vegetation in each of the strata are considered more complex. Simpler landscapes, characteristic of many newer residential developments, have limited or no overstory or understory, and an abundance of turfgrass. The degree of complexity within landscapes will affect the location and abun-dance of insects. For example, southern chinch bugs are pests that suck plant sap from grasses, mainly St. Augustine grass. It is widely observed that southern chinch bugs are more common in sunny locations (lacking overstory trees) in lawns compared to areas of the lawn shaded by small or large trees (*24,25*). Azalea lace bugs, a common pest of azaleas, are

FIGURE 2.2   A more complex urban landscape will have vegetation present in each layer or strata—overstory, understory, shrub layer, annual and perennial layer, and groundcover. Simpler landscapes will have limited or no overstory or understory.

another example. Similar to southern chinch bugs, it is common to find azaleas with greater damage from azalea lace bugs planted in full sun all day or in the afternoon (*26*). Plants with less damage from azalea lace bugs are planted in shaded locations with lower (<4000 footcandles) midday light intensity (*27*). While light intensity is a good predictor of damage, light intensity in landscapes can depend on the surrounding vegetation. Shrewsbury and Raupp (*23*) surveyed 24 residential and municipal land-scapes in Maryland and scored light intensity and structural complexity at each location. Within those urban landscapes, the tree layer and the presence of groundcover/turf layer accounted for most (76%) of the local abundance of azalea lace bugs. The presence of overstory trees, which would create shading, was negatively correlated with azalea lace bug populations. Likewise, more turfgrass which indicates limited or few mature trees and higher light, was positively correlated with azalea lace bug abundance. Furthermore, complex landscapes had greater abundance and retention of arthropod natural enemies (spiders and predatory insects) (*23*). So, landscapes with more turfgrass cover and fewer overstory trees (sunnier and less complex locations) would have more azalea lace bugs and those with more overstory trees (shadier and more complex locations) would have few azalea lace bugs. This type of information is typically only well documented for the most severe pests (chinch bugs and azalea lace bugs) in urban landscapes. Now that you are aware of these relationships, look for patterns in urban landscapes that you manage or design. Similar studies with other pests would improve our ability to better predict outbreaks and avoid those pitfalls during design.

> **BOX 2.1 Common categories used to define urban spaces.**
>
> Urbanization is not absolute such that one location is urbanized and another is not. Rather, a site can be categorized usually by the amount of impervious surfaces:
>
> **Rural or low urbanization**: Locations with <20% impervious surfaces.
>
> **Exurban**: An intermediate location linking the suburbs to rural areas. Typically, homesites with larger estate type lots.
>
> **Suburban or medium urbanization**: Locations having moderately (20%—50%) impervious surfaces due to housing and retail developments.
>
> **Urban core, downtown, inner city, or high urbanization**: More than 50% impervious surfaces.
>
> There are many studies showing the influence of the urban to rural gradient on insects. Raupp et al. (28) provides a great summary of that research.

It is not just the living components of urban landscapes that locally influence pests, but also the hardscape (concrete or paved surfaces). The hardscape within landscapes is typically measured by the amount (area or percent) of impervious, or water-resistant, surfaces. The amount of impervious surface is a way to categorize urbanization (see Box 2.1). Those categories are not absolute and higher quality research papers should define the criteria used to designate rural, suburban, or urban in the study. The proximity to, and the amount of, impervious surfaces influences urban landscape insects. Often greater pest populations will occur around host plants surrounded by more impervious surfaces (29). Impervious surfaces contribute structure, reduced habitat, and heat. As you may expect, tree temperatures in urban areas are correlated with impervious surfaces. In a series of papers, a research team at NC State University has carefully studied these factors relative to scale insects on urban trees. A lecanium (soft) scale insect on willow oak had 13 times greater populations in warmer urban areas in Raleigh, NC, than in cooler areas of the city. It is not just the temperature; the environment appears to change the scale insect populations. Scale insects collected from either warmer or cooler city landscapes perform differently when moved to trees on a different site. As temperatures warm, the scale insects from warmer city landscapes are better adapted to the heat (30). Similarly, populations of gloomy scale, an armored scale insect, on maple trees are closely correlated with greater tree temperatures (31). Since tree temperatures are correlated with the percent of impervious surfaces, they developed a novel way to estimate impervious surfaces and tree health called the Pace-to-Plant technique (32). The Pace-to-Plant technique is based on strong correlations between impervious surfaces and tree health along lines extending 10—100 m from the tree (Fig. 2.3). Within that range distance, 20 m (approximately 25 paces) is convenient because four lines of 25 paces is 100 total paces and easily converted to a percentage. An arborist or homeowner can take 25 paces away from the red maple tree along

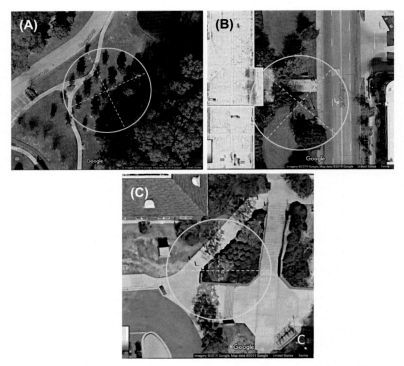

FIGURE 2.3   An application of the Pace-to-Plant technique on trees in Auburn, AL. Image A is a tree in good health; B is a tree in fair health; and C is a tree in poor health. Trees in Images B and C are infested with gloomy scale insects but not those in Image A. *Source: Google Maps.*

four equally spaced lines. Each step on impervious surfaces is counted giving a number out of 100 which is the estimate of the percent of impervious surfaces. If the number is 1%–32% impervious surfaces, then the maple tree is likely in good condition. When the percent is 32%–62% impervious surfaces, the tree is likely in fair condition. At 63% or more impervious surfaces, the red maple tree will be in poor condition (*32*). Fig. 2.3 shows trees in Auburn, AL, where this method was used, and corresponded well to trees that were good, fair, and poor when visually inspected. This tool could also be used to assess whether a particular site will be suitable for red maples (*32*) and produce more sustainable urban landscape plantings.

Relevance of spatial scales for management

Hopefully now you appreciate how spatial relationships can influence urban landscape pest and beneficial insects. Exotic species introductions provide some of the best examples where spatial relationships have been applied to pest management. The Emerald ash borer (EAB) is a wood-boring beetle introduced to North America from Asia (*33*). After being first detected in Michigan in 2001, it has spread into 31 states and 2 Canadian provinces. Natural spread of EAB is estimated at about 20 km/year. Females fly about 1 km per day and most eggs are found within a few 100 m away from where the adults emerge (*34*). This means that many host trees in a small town could have eggs within 1 year of adult emergence. For EAB, several studies have modeled the long-term sustainability of ash trees in urban forests and costs to local municipalities if ash were treated or removed. While the "knee jerk" reaction to detections of EAB by local municipalities is to remove trees, treating 20% of the trees in the entire urban forest annually would preserve almost all ash trees for 10 years. The insecticide costs over time would be less than the costs for tree removal (*35*). Given this, two additional studies (*36,37*) evaluated EAB management strategies using data from multiple municipalities. Both studies suggest a strong economy of scale effect, meaning lower average per tree treatment costs if more trees are treated. One study noted 44.7% lower costs if 150 or more trees are treated (*37*). Coordinated efforts across the borders of smaller cities should greatly improve EAB management and lower the costs over time compared to widespread tree removal (*36*).

Another exotic pest that demonstrates the importance of thinking spatially for management is the red imported fire ant, *Solenopsis invicta*. Imported fire ants deliver painful and sometime fatal stings to persons allergic to their venom. In the southeastern United States, urban landscapes are commonly treated to control red imported fire ants. Red imported fire ants are one of several related species of fire ants that entered the United States through the port of Mobile, Alabama. They have since spread to most southeastern and southwestern states (*38*). Imported fire ants are not easily managed locally. At a local scale, a typical urban home lot (0.25 acre) would have 15 or fewer mounds and homeowners would treat mounds individually four or more times per year (*38*). Managers of large tracts of land (e.g., plant nurseries or parks) within the fire ant range have specialized spreaders mounted to airplanes, trucks, or tractors to apply bait treatments (*38, personal observations*). Insecticidal baits, insecticides formulated on a food particle, have lower application rates and are efficiently applied at larger spatial scales, and can be overapplied when used at smaller spatial scales (*39*).

Several projects in Texas and Louisiana have demonstrated the effectiveness and costs associated with community-wide bait applications for imported fire ant control. There is also a planning tool for organizing these local community-wide fire ant programs (*40*). Most of these projects are not found in scientific journals because they only treat one subdivision (nonreplicated) and may not have comparisons with nontreated sites as controls. In some instances, homeowner feedback and control costs are provided to document adoption and the economy of scale for bait applications. There are some common outcomes across most or all of these projects (*41–46*): (1) there is high homeowner participation, (2) there are cost savings relative to nonbaited sites or sites not managed with baits, and (3) mound reductions or ant activity usually persists longer in community-wide management programs. Most programs have 1–2 years of data (*41,42,44–46*) with the exception of a 40.5 ha (100 acre) subdivision in Sante Fe, Texas, that has been under community-wide fire ant management for >15 years (*43*). Interestingly, the benefits of community-wide ant management may be unique to bait formulations and may not translate to other forms of control. A replicated study in Louisiana found that granular (not baits) insecticides produced similar mound reductions and duration of control up to 7 months after application regardless of the size of the treated area (*47*).

Like area-wide control with insecticides, biological control projects using insects also work well at broader spatial scales. Large spatial scales are useful when releasing new insects for biological control, and some natural enemies "follow" introduced pests as they expand their territory. California, Florida, and Hawaii have high introduction rates for nonnative insect species, and most of the well-documented examples with biological controls are in those states. Nonnative plants may have fewer insect herbivores (*48,49*), yet the introduction of an herbivore from the original "homeland" of the plant can be particularly devastating. One response by state and federal governments is to organize expeditions to the origin of those plants and pests, find natural enemies, then bring and release them into the expanded range. In the early 1900s many expeditions resulted in the unregulated release of dozens of nonnative insects into the United States for biological control of exotic pests like Japanese beetles (*7*). In many cases, most of those insects were never captured again and assumed to never have established. The process still occurs today but with far greater scrutiny from federal and state regulators.

Let us consider two examples where insects introduced for biological control have ignored the smaller town or city boundaries. The ash whitefly, a pest originally introduced into California, is now established in California and Florida (*50*). In California, scientists at the University of California and the California Department of Food and Agriculture located, mass produced, and released a small parasitoid wasp, *Encarsia partenopea*, for biological control. Within 5 months of the release, there was a 13-fold reduction in the number of whiteflies at the release sites. The parasitoid wasp was also being collected at nonrelease sites that were 11 km (7 miles) from the release sites (*51*). Similarly, from the late 1970s until the early 1990s, the state of Florida and the University of Florida's Institute of Food and Agriculture operated the Mole Cricket Research program with an emphasis on finding and releasing natural enemies of three species of nonnative mole crickets. *Larra bicolor*, a parasitoid wasp (Fig. 2.4), was among those natural enemies imported and released. In Florida, the wasp along with other biological control agents released is attributed with significant reductions in mole cricket populations (*52*). Interestingly, there are no records of this wasp being imported or released into

FIGURE 2.4   An adult *Larra bicolor* feeding on flowers. This parasitoid wasp was successfully imported and released into Florida for biological control of mole crickets and has since spread to other states in the southeastern United States.

Mississippi, Alabama, or North Carolina, yet populations of the wasps seemingly spread coincident with mole crickets into those states (*53–55*).

State or federal insect biological control programs often operate without much public recognition or intervention. It has been argued (*56*) that not engaging the general public is a missed opportunity by biological control programs. The ash whitefly control project in California was among the first that I can recall to tap the power of the general public. After demonstrating that the parasitoid could effectively reduce ash whitefly populations, the University of California and the California Cooperative Extension began offering parasitoids in shipments of 250 wasps to the public for release locally. The wasps were available to order and sold to only recover rearing costs. The majority of orders came from private citizens some of which collected orders for neighborhood-wide distribution. This program released 120,000 parasitoids throughout the state, arguably a larger geographical distribution than could be reached by a single university or state government program alone (*56*).

Social media and crowdsource funding should make engaging the general public in these programs easier than in the past. But, there have been few, if any, urban landscape biological control programs that have matched the level of public engagement and impacts of the ash whitefly project. In fact, in 2017, the US Department of Agriculture cut funding for the phorid fly rearing programs leaving states with established, yet isolated populations of phorid flies (*57*). The continued involvement and financial investment of private citizens are common themes among the successful community-wide management and biological control programs highlighted here. Future programs modeled after these successes could be one avenue where future Extension specialists in urban landscapes could make significant impacts.

Highlights

- Urban insects in landscapes often live and behave in spatial scales outside of the constraints of the individual landscape where they are managed.
- There are some predictable relationships between plant and insect diversity and location within a city (inner city, suburban, rural).
- Despite our existing and growing knowledge about these spatial relationships, few pests, except a few invasive species, are managed at an appropriate spatial scale.
- Successful area-wide management programs with significant public engagement are models on which to develop future projects.

References

1. R.H. MacArthur, E.O. Wilson, The Theory of Island Biogeography, Princeton University Press, Princeton NJ.
2. Fattorini, S.; Mantoni, C.; de Simoni, L.; Galassi, D. M. P. Island Biogeography of Insect Conservation in Urban Green Spaces. *Environmental Conservation* **2018**, *45* (1), 1−10.
3. Gibb, H.; Hochuli, D. F. Habitat Fragmentation in an Urban Environment: Large and Small Fragments Support Differnt Arthropod Assemblages. *Biological Conservation* **2002**, *106*, 91−100.
4. Letheren, A.; Hill, S.; Salie, J.; Parkman, J.; Chen, J. A Little Bug with a Big Bite: Impact of Hemlock Woolly Adelgid Infestations on Forest Ecosystems in the Eastern USA and Potential Control Strategies. *International Journal of Environmental Research and Public Health* **2017**, *14*, 438. https://doi.org/10.3390/ijerph14040438.
5. Lockwood, J. A. Voices from the Past: What We Can Learn from the Rocky Mountain Locust. *American Entomologist* **2001**, *47* (4), 208−215.
6. Eckert, J. E. The Flight Range of the Honeybee. *Journal of Apicultural Research* **1933**, *47* (5), 257−285.
7. Fleming, W. E. *Biology of the Japanese Beetle, Tech. Bull. No. 1449;* US Department of Agriculture: Washington, 1972.
8. Showers, W. B. Migratory Ecology of the Black Cutworm. *Annual Review of Entomology* **1997**, *42*, 393−425.
9. Nagoshi, R. N.; Meagher, R. L.; Hay-Roe, M. Inferring the Annual Migration Patterns of Fall Armyworm (Lepidoptera: Noctuidae) in the United States from Mitochondrial Haplotypes. *Ecol Evol* **2012**, *2* (7), 1458−1467.
10. Medler, J. T. Migration of the Potato Leafhopper-A Report on a Cooperative Study. *Journal of Economic Entomology* **1957**, *50* (4), 493−497.
11. Magsig-Castillo, J.; Morse, J. G.; Walker, G. P.; Bi, L. B.; Rugman-Jones, P. F.; Stouthamer, R. Phoretic Dispersal of Armored Scale Crawlers (Hemiptera: Diaspididae). *Journal of Economic Entomology* **2010**, *103* (4), 1172−1179.
12. Wainhouse, D. Dispersal of First Instar Larvae of the Felted Beech Scale, *Cryptococcus fagisuga. Journal of Applied Ecology* **1980**, *17* (3), 523−532.

13. Rabkin, F. B.; Lejeune, R. R. Some Aspects of the Biology and Dispersal of the Pine Tortoise Scale, *Toumeyella numismaticum* (Pettit and McDaniel) (Homoptera: Coccidae). *The Canadian Entomologist* **1954**, *86* (12), 570–575.

14. Wegner, G. S.; Niemczyk, H. D. Bionomics and Phenology of *Ataenius spretulus*. *Annals of the Entomological Society of America* **1981**, *74*, 374–384.

15. Diaz, M. D. C.; Peck, D. C. Overwintering of Annual Bluegrass Weevils, *Listronotus maculicolis*, in the Golf Course Landscape. *Entomologia Experimentalis et Applicata* **2007**, *125*, 259–268.

16. Kerr, S. H. Biology of the Lawn Chinch Bug, *Blissus insularis*. *The Florida Entomologist* **1966**, *49* (1), 9–18.

17. Rieske, L. K.; Townsend, L. H. Orientation and Dispersal Patterns of the Eastern Tent Caterpillar, *Malacosoma Americanum* F. (Lepidoptera: Lasiocampidae). *Journal of Insect Behavior* **2005**, *18* (2), 193–207.

18. Rhainds, M.; Gries, G.; Ho, C. T.; Chew, P. S. Dispersal by Bagworm Larvae, *Metisa plana:* Effects of Population Density, Larval Sex, and Host Plant Attributes. *Ecological Entomology* **2002**, *27*, 204–212.

19. Ghent, A. W. Studies of Ballooning and Resulting Patterns of Locally Contagious Distribution of the Bagworm *Thyridopteryx ephemeraeformis* (Haworth) (Lepidoptera:Psychidae). *The American Midland Naturalist* **1999**, *142* (2), 291–313.

20. Cox, D. L.; Potter, D. A. Aerial Dispersal Behavior of the Bagworm. *Journal of Arboriculture* **1990**, *1*6 (9), 242–243.

21. Williamson, R. C.; Potter, D. A. Nocturnal Activity and Movement of Black Cutworms (Lepidoptera: Noctuidae) and Response to Cultural Manipulations on Golf Course Putting Greens. *Journal of Economic Entomology* **1997**, *90* (5), 1283–1289.

22. Markin, G. P.; Dillier, J. H.; Hill, S. O.; Blum, M. S.; Herman, H. R. Nuptial Flight and Flight Ranges of the Imported Fire Ant. *Journal of Entomological Science* **1971**, *6*, 145–156.

23. Shrewsbury, P. M.; Raupp, M. J. Evaluation of Components of Vegetational Texture for Predicting Azalea Lace Bug, *Stephanitis pyrioides* (Heteroptera: Tingidae), Abundance in Managed Landscapes. *Environmental Entomology* **2000**, *29* (5), 919–926.

24. Reinert, J. A.; Kerr, S. H. Bionomics and Control of Lawn Chinch Bugs. *Bulletin of the Entomological Society of America* **1973**, *19*, 91–92.

25. Kaur, N.; Gillett-Kaufman, J. L.; Gezan, S. A.; Buss, E. A. Association between *Blissus insularis* Densitis and St Augustinegrass Lawn Parameters in Florida. *Crop Forage Turfgrass Manage* **2016**, *2*. https://doi.org/10.2134/cftm2016.0015.

26. Raupp, M. J. Effects of Exposure to Sun on the Frequency of Attack by the Azalea Lacebug. *Journal of American Rhododendron Society* **1984**, *38* (4), 189–190.

27. Trumbule, R. B.; Denno, R. F. Light Intensity, Host-Plant Irrigation, and Habitat-Related Mortality as Determinants of the Abundance of Azalea Aace Bug (Heteroptera: Tingidae). *Environmental Entomology* **1995**, *24* (4), 898–908.

28. Raupp, M. J.; Shrewsbury, P. M.; Herms, D. A. Disasters by Design: Outbreaks along Urban Gradients. In *Insect Outbreaks Revisited;* Barbosa, P., Letourneau, D. K., Agrawal, A. A., Eds.; Wiley-Blackwell: UK, 2012; pp 313–340.

29. Raupp, M. J.; Shrewsbury, P. M.; Herms, D. A. Ecology of Herbivorous Arthropods in Urban Landscapes. *Annual Review of Entomology* **2010**, *55*, 19–38.

30. Meinke, E. K.; Dunn, R. R.; Sexton, J. O.; Frank, S. D. Urban Warming Drives Insect Abundance on Street Trees. *PLoS One* **2013**, *8* (3), e59687. https://doi.org/10.1371/journal.pone.0059687.

31. Dale, A. G.; Frank, S. D. The Effects of Urban Warming on Herbivore Abundance and Street Tree Condition. *PLoS One* **2014,** *9* (7), e102996. https://doi.org/10.1371/journal.pone.0102996.

32. Dale, A. G.; Youngsteadt, E.; Frank, S. D. Forecasting the Effects of Heat and Pests on Urban Trees: Impervious Surface Thresholds and the 'Pace to Plant' Technique. *Arboriculture & Urban Forestry* **2016,** *42* (3), 181—191.

33. *Emerald Ash Borer Information Network,* 2018. U.S.A. http://www.emeraldashborer.info/.

34. Valenta, V.; Moser, D.; Kapeller, S.; Essl, F. A New Forest Pest in Europe: A Review of Emerald Ash Borer (*Agrilus planipennis*) Invasion. *Journal of Applied Entomology* **2016,** *141,* 507—526.

35. McCullough, D. G.; Mercader, R. J. Evaluation of Potential Strategies to SLow Ash Mortality (SLAM) Caused by Emerald Ash Borer (*Agrilus planipennis*): SLAM in an Urban Forest. *International Journal of Pest Management* **2012,** *58* (1), 9—23.

36. Kovacs, K. F.; Haight, R. G.; Mercader, R. J.; McCullough, D. G. A Bioeconomic Analysis of an Emerald Ash Borer Invasive of an Urban Forest with Multiple Jurisdictions. *Resource and Energy Economics* **2014,** *36,* 270—289.

37. Sadof, C. S.; Hughes, G. P.; Witte, A. R.; Peterson, D. J.; Ginzel, M. D. Tools for Staging and Managing Emerald Ash Borer in the Urban Forest. *Arboriculture & Urban Forestry* **2017,** *43* (1), 15—26.

38. Drees, B. M.; Calixto, A. A.; Nester, P. R. Integrated Pest Management Concepts for Red Imported Fire Ants *Solenopsis invicta* (Hymenoptera: Formicidae). *Insect Science* **2013,** *20,* 429—438.

39. Nester, P. R.; Puckett, R. T.; Flanders, K.; Palmer, K.; Graham, F.; Suiter, D.; David, T.; Vail, K.; Hopkins, J.; Loftin, K.; Ring, D. Broadcast Baits for Fire Ant Control. eXtension https://articles.extension.org/pages/74558/broadcast-baits-for-fire-ant-control.

40. Nester, P.R.; Brown, W. Community-wide Imported Fire Ant Management Kit. https://fireant.tamu.edu/files/2014/03/ENTO_025.pdf.

41. Hooper-Bùi, L. M.; Pranschke, A. M.; Story, H. M. Area-wide Bait Treatments Compared with Individual Homeowner Treatments for Red Imported Fire Ant Management. In *Proceedings of the 2000 Annual Imported Fire Ant Conference;* 2000; pp 71—74.

42. Nester, P. R. "Fighting Texas' Fire Ants: the Team Approach" 15 Years and Counting-Review of the Lago Santa Fe Fire Ant Project. In *Proceedings of the 2017 Annual Imported Fire Ant Conference;* 2017; pp 18—24.

43. Barr, C. L.; Best, R. L.; McCarver, R. G. In *Proceedings of the 2000 Annual Imported Fire Ant Conference;* 2000; pp 47—52.

44. Riggs, N. A Tale of Two Neighborhoods: Examination of Fire Ant Infestations before and after Community-wide Fire Ant Treatments in Two San Antonio, Texas Neighborhoods. In *Proceedings of the 2000 Annual Imported Fire Ant Conference;* 2000; pp 53—58.

45. Russell, S. A. Community-wide Fire Ant Management in Dallas and Tarrant Counties, Texas: An Overview. In *Proceedings of the 2000 Annual Imported Fire Ant Conference;* 2000; pp 59—60.

46. Lennon, L. Travis. Williamson Counties' Showcase Programs: An Update. In *Proceedings of the 2000 Annual Imported Fire Ant Conference;* 2000; pp 61—66.

47. Hooper-Bùi, L. M.; Story-Whitney, H.; Legendre, J. Management of Red Imported Fire Ants on Large-Scale and Small Scale. In *Proceedings of the 2001 Annual Imported Fire Ant Conference;* 2001; pp 72—73.

48. Tallamy, D. W.; Shropshire, K. J. Ranking Lepidopteran Use of Native versus Introduced Plants. *Conservation Biology* **2009,** *23* (4), 941—947.

49. Clem, S. C.; Held, D. W. Species Richness of Eruciform Larvae Associated with Native and Alien Plants in the Southeastern United States. *Journal of Insect Conservation* **2015**, *19,* 987—997.

50. Nguyen, R.; Hamon, A. B. *Ash Whitefly, EENY-147;* University of Florida, 2017.

51. Bellows, T. S.; Paine, T. D.; Gould, J. R.; Bezark, L. G.; Ball, J. C.; Bentley, W.; Coviello, R.; Downer, J.; Elam, P.; Flaherty, D.; Gouveia, P.; Koehler, C.; Molinar, R.; O'Connell, N.; Perry, E.; Vogel, G. Biological Control of Ash Whitefly: a Success in Progress. *California Agriculture* **1992,** *46* (1), 27—28, 24.

52. Frank, J. H.; Walker, T. J. Permanent Control of Pest Mole Crickets (Orthoptera: Gryllo-talpidae: *Scapteriscus*) in Florida. *American Entomologist* **2006,** *52,* 138—144.

53. Held, D. W. Occurrence of *Larra bicolor* (Hymenoptera: Sphecidae), Ectoparasite of Mole Crickets (*Scapteriscus* spp.), in Coastal Mississippi. *The Florida Entomologist* **2005,** *88,* 327—328.

54. Abraham, C. M.; Held, D. W.; Wheeler, C. First Report of *Larra bicolor* (Hymenoptera: Sphecidae) in Alabama. *Midsouth Entomologist* **2008,** *1,* 81—84.

55. Hertl, P. T.; Brandenburg, R. L. First Record of *Larra bicolor* (Hymenoptera: Crabronidae) in North Carolina. *The Florida Entomologist* **2013,** *96* (3), 1175—1176.

56. Paine, T. D.; Millar, J. G.; Bellows, T. S.; Hanks, L. M. Enlisting an Underappreciated Clientele: Public Participation in Distribution and Evaluation of Natural Enemies in Urban Landscapes. *American Entomologist* **1997,** *43,* 163—173.

57. Callcott, A. M. The USDA-APHIS Phorid Fly Release Project: 2001-2017. In *Proceedings of the 2017 Annual Imported Fire Ant Conference;* 2017; pp 33—36.

Chapter 3

Plants as active participants in urban landscapes

Immigrants, bad neighbors: these are not just topics you might discuss during Happy Hour with friends. These are terms relevant to a discussion about plants in urban landscapes. Plants are the most conspicuous biological components of urban/suburban landscapes. In Chapter 1, we learned that plants, as the primary food and habitat source for insects, are important shapers of insect communities. Plants harvest the energy from the sun and translate that energy into a biomass that delivers mainly nitrogen and carbon to herbivorous animals. In this chapter, you will learn some basic concepts of plant-insect interactions, plant diversity, esthetics, and the functions of plants in urban landscapes. These concepts are useful in understanding the role of plants in pest outbreaks and the ecology driving the native plant movement.

Introduction to insect-plant relationships

People that focus on the sale and maintenance of plants perceive herbivorous insects and mites and their damage more than other (predators or parasitoids) groups of arthropods. In this context, you might conclude, albeit incorrectly, that most insects use plants as food. If you look across the diversity of insect species (millions of species), most are not herbivores. *Since plants have the greatest biomass on the planet, why is plant feeding not more common?* Because plants are not a good source of nitrogen (Fig. 3.1). Plant parts are high in water content (60%–80%) but low in nitrogen content. Nitrogen is often a limiting resource for insects. Considering the food items in urban landscapes available for insects, the best choices for nitrogen would be insects, seeds, or fungi. Natural enemies and scavengers eat insects or insect parts. Ambrosia beetles, such as the granulate ambrosia beetle (*Xylosandrus crassiusculus*), consume fungi growing inside the host tree rather than the tree. This makes ecological sense because ambrosia fungi are a better source of nitrogen for the developing larvae.

Among plant parts, there is still tremendous variation in food quality (Fig. 3.1). Leaves of most plant groups have the highest nitrogen content followed by roots and stems, with xylem at the bottom. Xylem is the plant sap that is essentially mineral water. These food choices of insects can help explain

Urban Landscape Entomology. https://doi.org/10.1016/B978-0-12-813071-1.00003-8

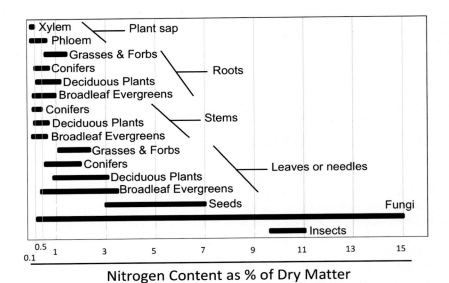

FIGURE 3.1 The relative nitrogen content of insect foods in nature (*1,2,3*).

why some insects develop faster than others. Most white grubs that consume grass root have one generation per year. Cicadas that consume xylem take 2 or more years to develop through their immature stages. Contrast that life history to the multiple generations of virtually all predatory insects and spiders. Plant parts may be widely available, but those insects or mites using plants have consequences for growth and development of immatures, or life span and reproduction in adult insects.

Insect herbivores are largely restricted to certain plants and plant parts. For example, azalea caterpillars and azalea lace bugs feed only on leaves, and white grubs feed only on roots. Leafminers consume leaves, but only from the inside. This specificity in feeding is helpful in diagnosing pest problems. Most insect herbivores are plant specialists, only consuming plants in a few families or plant in only one genus (e.g., milkweeds and Monarch larvae). This, "botanical instinct" as it was once thought to be, is not due to extensive lessons in horticulture, but expertise in organic chemistry. Yes, organic chemistry; the course most college students dread. Yet, insects specialize in it. Primary nutrients such as nitrogen in plants can affect plant (host) quality for insects. But, it is the organic chemicals called secondary plant compounds that help insects recognize their host plants from among other plants and nonplant surfaces. Insects have tiny receptors on their antennae, legs, and mouthparts that are used to sample food. These external receptors mean that insects do not need to rely solely on taste through bites to sample a plant. This makes ecological sense because secondary plant compounds are very toxic to insects that lack adaptations to deal with them. A few bites may be enough for an insect to be impaired, get sick, or even die. These secondary plant compounds are an important factor used to determine host range or the breadth of plant species or genera, for insects. Many specialist insect herbivores have sensory receptors that only turn on, like a switch, when they detect a certain amount of one or two specific secondary plant compounds. Imagine if you could wiggle your fingers or toes into your food and know if you are being served asparagus or kale. That is essentially what these insects are doing. This circuitry in specialist insects may allow them to make more efficient decisions than dietary generalist insects that feed on a wide range of host plants (4).

Generalist insect herbivores, such as Japanese beetles or gypsy moths, feed on many unrelated plants. These generalists, especially when introduced from elsewhere, often excel at being pests (e.g., Japanese beetles, gypsy moths) because they have a wide range of available host plants. Because a generalist often requires tasting or a partial meal before making the decision to stay or go, they can consume plants that are edible, but toxic (Fig. 3.2). These differences in host range can be useful in diagnostics and problem solving. For example, an azalea caterpillar on azaleas will not damage adjacent roses or gardenia bushes because those plants are not hosts for that species. With this information, you would know that treatment of other plants that were not azaleas in the landscape would be a waste of time and overuse of insecticides. While it would be nice to list here the host ranges for all major insect pests in urban landscapes, there are too many insects and host plants. Partial or complete host records for common urban landscape pests are widely available in books (6,7), online resources such as the caterpillar database (8), or University Extension websites (9). These diagnoses do require some background knowledge of plant and insect identification, which we will learn in later chapters.

FIGURE 3.2   Adult Japanese beetles are generalist herbivores consuming >300 plant species in more than 70 plant families. That range of food can lead to poor choices like this one that is "drunk" from consuming geranium flowers (5).

FIGURE 3.3 Mole cricket nymphs fed a plant or animal diet, or a mixture of plant and animal food for 2 months. The two larger insects on the left consumed insects (as part of their diet), but insect protein was excluded from the smaller mole cricket on the right. This shows the benefit of consuming food with a greater nitrogen content like animal tissues (*10*).

Finally, there is an unusual group of insects called omnivores, which are not restricted to a particular food. Mole crickets are common pests in turfgrass especially in the sandy soils in the Gulf States. Tawny and southern mole crickets are omnivores, consuming plant and animal tissues. In lab studies, mole crickets fed insects develop faster than those fed plants (*10*). The mole crickets in Fig. 3.3 all hatched from the eggs at about 2 months before the image was taken. Mole crickets that consumed insects (animal or rotation) were fifth and sixth instar nymphs compared with the smaller second instar nymphs provided a plant-only diet. Mole crickets can consume plants, which are widely abundant in their habitat. But they will benefit developmentally when the opportunity arrives to eat a smaller mole cricket or other insects.

Plant density and diversity

In this section, we will use data on trees in urban green spaces (private and public landscapes) and the definitions for the degrees of urbanization from Chapter 2. First let us consider the urban core. Intuitively, trees cannot grow in concrete, so tree density is well correlated with the amount of pervious (water permeable) surfaces (*11,12*). As more land is cleared and paved, less space exists for the root volume of urban trees. An exception are arid, desert locations (Arizona, Nevada, and parts of California) where supplemental irrigation is installed. In those locations, urbanization may increase overall tree density.

The majority of the literature, however, is spent discussing changes in plant diversity in relation to urbanization. There is a strong relationship between plant diversity with the degrees of urbanization (*13*). You may assume, as I once did, that rural areas, with remnant stands of native plants, would have the most plant diversity. However, among the 17 studies reviewed by McKinney (*13*), most show an increase in species richness in suburban areas compared with rural. If you rank the degrees of urbanization for plant diversity, it would be suburban > rural > urban (inner city). The lack of diversity in rural areas is primarily driven by agricultural, pasture, and forest lands, which have low diversity and high uniformity over relatively large tracts of land. The following sections will summarize what is known about suburban plant diversity and the effects it has on insect-plant relationships.

Plant diversity may be more easily explained in terms of agriculture. If you have a large agricultural field of one crop species (corn, soybeans, hops, rice), it is a monoculture. Monocultures are stands of plants with low genetic diversity and high uniformity. The opposite end of the continuum, called polycultures, has great genetic diversity and low uniformity. Based on the previous section, you may assume suburban landscapes are mainly polycultures, and you would not be incorrect. However, sometimes, plant components in these landscapes are mass planted and essentially large monocultures. For example, the Auburn University campus has >1600 crapemyrtle trees on campus, about three times more than the next most used tree (*14*). That is 2.2 trees per ha (0.9 trees per acre) based on total land area not just green space. Monoculture is really about low diversity at high density, but the actual density that makes something a monoculture is not well defined. This is relevant because mass plantings and monocultures influence insect as much as polyculture. Monocultures tend to accumulate specialist herbivores, either they are more efficient at finding those patches or they find the patch and just do not leave (*15*). In polycultures, there are generally less specialists and more generalists. While these concepts were developed mostly from studies in agricultural fields, they are still relevant to urban landscapes.

Street tree plantings and home lawns are two prominent examples of monocultures in urban landscapes. Street trees are usually the same species or cultivar of tree planted in rows equally spaced from one another to emphasize uniformity (Fig. 3.4). The diversity of street trees can even carry over between adjacent communities (*16*). In the southeastern and southwestern United States, many communities have one or more main streets lined with crapemyrtle trees (*Lagerstroemia* spp.), known as the "tree of the south." Urban street trees, species such as crapemyrtle, ash (*Fraxinus* spp.), elm (*Ulmus* spp.), or maples (*Acer* spp.) are often identified as being successful street trees and then get extensively planted. In a survey of street trees in 12 US cities, 32 genera of street trees were used, but maples (*Acer*) accounted for as much as 57% of the street trees in those surveyed cities (*17*). Other street tree surveys (*18,19*) reveal considerable similarity among genera of trees (*Quercus, Tilia, Prunus, Robina, Ulmus, Populus,* and *Acer*) used worldwide in street plantings. The monotonous planting of trees is comparable with pioneer plants in ecological succession (*11*). Pioneer plants usually do well in poor growing conditions and disturbed environments. Large blocks of related trees have historically been susceptible to exotic pests such as Emerald ash borer and crapemyrtle bark scale that have reduced or eliminated their utility as street

FIGURE 3.4 Suburban landscapes are more diverse than urban or rural landscapes. However, street tree plantings and lawns are still situations with little overall plant diversity.

trees (*17*). The low diversity in street trees contrasts sharply with the greater diversity of trees in residential landscapes within the same community (*16*).

Urban lawns are another example of monocultures in urban landscapes. Turfgrass is the largest irrigated crop in the United States (*20*) and a significant feature in urban residential and institutional landscapes. Lawns occupy as much as 20%−30% of the land cover in urban areas and are common worldwide (*21,22*) despite being considered an exclusive feature of American suburban landscapes. Urban citizens worldwide value lawns for various social and recreational activities. Ignatieva et al. (*22*) provide one of the best studies on the use patterns and perceptions of lawns. They conclude, however, that not all urban citizens apparently value the current, monotonous version of lawns (*22*), a perspective resonating in recent years with urban citizens (*23*). Lawns and out-of-play rough areas on golf courses, composed of grasses intermixed either with broadleaf annual and perennial plants, provide habitats that enhance local urban biodiversity (*24−27*). Within the United States, meadowlike lawns, however, can draw concerns from homeowner associations that seek a "green standard" for home lots in their communities. Furthermore, pests can still outbreak in a lawn mixed with grass and weeds. In those situations, the application method, cultural approaches, and the selection of insecticides are important consideration to reduce risks to pollinators and beneficial insects (*28−31*).

At this point in a class, I would ask students if we should increase plant diversity in landscapes to reduce pest problems, or does the specific mix of plants matter? Increasing plant diversity is nice for esthetics but may be counterproductive when used as a strategy to reduce pests. Raupp et al. (*32*) evaluated the influence of plant diversity or abundance in 200 urban landscapes over 2 years in Maryland. Plant abundance and plant diversity are good predictors of the number of pest species in urban landscapes. When the relationship between pest species and plant diversity and abundance are plotted, pest species accumulate more readily in landscapes with the addition of new plant species than the addition of more of the same species. The addition of new plant species provides better opportunities to add new pest species than increasing abundance of ornamental plants already present (32). While that study suggests that pest species richness could increase, it does not indicate if pests are more abundant. In another field study (33), flower beds with two roses were interplanted with three companion plants, some alleged to repel or deter insects. The number of Japanese beetles on roses interplanted was similar to that of roses alone in the flower beds. Roses interplanted with zonal geraniums had more beetles on the interplanted roses than roses planted alone (33). Interplanting host plants with other plants may cause specialist insects to land more readily on nonhost plants (i.e., inappropriate landings hypothesis) and leave (34). The addition of plant species to lawns or land-scapes may have great esthetic or social value, but this diversity *per se*, for the reduction of pests, is not well supported. As we will see in later sections, the addition of certain plants to landscapes can influence (positively and nega-tively) neighboring plants and the types and abundance of natural enemies present.

Earlier, we noted that suburban landscapes support more plant diversity than urban or rural areas; however, woody plant diversity in urban landscapes is narrowly focused in a few key plant families and genera. Two plant families, Rosaceae and Pinaceae, are overrepresented in urban landscapes (35−37). Pine and rosaceous plants have been widely used for food and fiber crops in temperate regions including the United States for thousands of years (38). Almost all rosaceous plants in landscapes (crabapples, roses, ornamental cherry) are commonly attacked by herbivorous mites and insects whenever present (35,39). If you recall, herbivorous mites and insects are experts in plant taxonomy and organic chemistry. The relatedness and abundance of rosaceous plants and pines across different landscapes makes it easy for herbivores to find those plants and succeed. For example, aphids or caterpillars that specialize on rosaceous plants should have no problem finding and using crabapple or landscape roses installed in a new landscape. Urban landscapes in the United States are also dominated by about 20 plant genera. Oak (*Quercus* spp.), maple (*Acer* spp.), holly (*Ilex* spp.), and azaleas (*Rhododendron* spp.) are represented over a large geographic range (35,36). Common taxa in landscapes have been called "key plants" because they contribute to the pleasure of the property through either esthetics or function (natural screens or shade). Interestingly, surveys suggest the 20 most commonly planted genera of plants in landscapes account for 77%−97% of the problems with insect and mite pests in landscapes (39).

Native or nonnative: does it really matter?

Native plants contribute to a sense of place. In the United States, Savannah Georgia and Biloxi Mississippi have large coastal live oak trees with Spanish moss hanging from their branches. These two native plants like the large sequoias of northern California provide context and an identity to those cities or regions. Urbanization typically reduces stands of native trees to small remnant populations surrounded by nonnative plants. If that occurs, it could change the identity of those locations. Furthermore, those changes would have effects on the associated communities of insects and ecosystem services.

Novelty, vigor, and year-long esthetic and visual appeal (flowering, fall color, shade) are reasons for the reliance on exotic (nonnative) plants in urban landscapes. For example, crapemyrtles do well in the dry and warm southern climate and have beautiful floral displays at times when few other plants are flowering. Crapemyrtles are native to Asia and just one example of how urbanization is an important gateway to exotic plant species. Across field surveys of urban landscapes and parks, most (45%−70%) of the plant diversity is represented by exotic plants (*12,16,37,40*). Nonnative species also outnumber native trees species used for street trees or in urban forests (*14,18*). For example, in New York state, the nonnative Norway maple is the most abundant street tree (*19*), and crapemyrtle is the most abundant tree on the Auburn University campus (*14*). Interestingly, older landscapes often have more vegetation than new landscapes (*16*) but a higher percent of nonnative plants (*37*). Exotics or even invasive plants, and not native plants, are being introduced accidentally or deliberately over time.

The nursery industries in the United States and abroad are common promoters of exotic or limited plant choices. Inventories in nurseries represent mostly exotic plants that are easy to vegetatively propagate at a low cost (*37,41*) or plants perceived as being locally adapted. Native plants may be more difficult to grow, or nurseries may want to sell trademarked varieties that have name recognition from online and in-print promotions. In their survey of US nurseries across four mid-Atlantic states, Coombs and Gilchrist (*41*) noted 24.7% of plant taxa available for sale were native, and 2% (26 species) were deemed invasive. Using plant catalogs, researchers determined that plant diversity in nurseries near Los Angeles County, California, is largely driven by exotic species (*42*). This has national implications because this area in California is one of the larger plant-producing areas in the United States, supplying plants to regional and national retail outlets. Homeowners and urban foresters can only use plants available to locally or regionally in the trade. For these reasons, the plant production industries are important stakeholders in discussions about plant diversity and ecosystem services in urban landscapes.

Studies (*43−47*) have documented the responses of insects to native and exotic plants. Exotic plants with no closely related (in the same genus) plants in an area will often escape attack by herbivores in areas where they are not native. The inverse is also true, native plants commonly host a greater number of species of insects than exotic plants. Crapemyrtle, *Ginkgo biloba*, and privet (*Ligustrum* spp.) are examples of exotic plant genera not present in the United States before they were introduced. In the southeastern United States, crapemyrtle and ginkgo both have four species of generalist caterpillars that can feed on their foliage. Other exotic plants are considered congeners (same genus) because they have a related plant already in an area where they are being introduced. In the United States, Norway maple (*Acer platanoides*) is congeneric with native sugar and red maple. Because there was a known relative in the United States, some insect herbivores can be "shared" between these species. The 11 species of caterpillars that feed on Norway maple in the southeastern United States were likely using native maple trees as hosts before Norway maple was introduced (*47*).

This idea that exotic plants experience less attacks in a new area is the enemy release hypothesis (*45*). The same principle applies to exotic insects. An organism relocated to a new area will have less natural competition or fewer attacks from predators, giving it an advantage in the new location. This is one of several routes by which a plant, disease, or insect can become a pest. In addition to esthetic reasons, resistance to pests may be another reason why the plant industries introduce exotic plants. After all, host plant resistance is an important tool in integrated pest management (IPM) and will be discussed in detail in Chapter 8. History proves, however, that the enemy escape experienced by these exotic plants is not long-lived. Eucalyptus trees, for example, were moved from their native range in Australia to southern Africa, North America, and South America. These trees had about 100 years of limited pest pressure in the United States before several specialist pests arrived in California in the 1980s (*48*). The global movement of plants has also been a source of several exotic pest introductions to North America such as Japanese beetles and other scarab beetles, Gypsy moths and several other caterpillars, and scale insects.

In fact, the iconic cherry trees around the Tidal Basin in the United States capital and the focus of the National Cherry Blossom Festival are a story of exotic plants and exotic scale insects (*50*). The first shipment of 2000 cherry trees donated on behalf of the citizens of Tokyo was inspected by a team of zealous USDA inspectors. Upon inspection, this shipment was infested with multiple exotic scale insects and crown gall disease (*Agrobacterium tumefaciens*) and then burned in a large pile on the grounds of the Washington Monument (*50*)! Another shipment of trees was sent, and those eventually became the original planting at the Tidal Basin.

To be fair, exotic plants are important sources of food for humans and wildlife, fiber, and esthetic value. Corn, rice, soybean, peaches, tomatoes (and others) are not native to the United States, but they are important annual and perennial food plants. There are ongoing discussions about the value of exotic and native plants for insects in urban landscapes (see Box 3.1), mainly focused on ecosystem services. As stated earlier, exotic plants host fewer insects,

BOX 3.1 Are the impacts of exotic plants overstated?

Remember, the Jenga® tower represented an existing community or food web. As with the tower, components in a food web support one another. As we remove and replace plants, it is likely that they will not fit exactly into the tower as the original plant. The degree of change (entirely new genus or just a different species) will contribute to the impacts of the change.

Because insects are important food for birds and other predators, reducing insects in urban landscapes can affect other animals (secondary consumers) that rely on insects (49). This is one of the discussion points about the roles of exotic and native plants in urban landscapes.

- One side argues that the use of exotic plants destabilizes the entire food web to a point of collapse or near collapse.
- Another side argues that urban landscapes are already so disturbed that one more disturbance is like trying to see ripples when you throw a rock into a crashing wave.

which can result in fewer insecticide applications during production and in the landscape. These are positives and consistent with IPM and environmental stewardship. Exotic plants also function as nectar and pollen resources for honey bees and native bees. Among 72 woody plants species growing in urban and suburban sites, Mach and Potter (51) did not find a difference in bee diversity or abundance (flower visits) between native and nonnative plants. Exotic plants may provide floral resources at times when native plants are not flowering (51,52). For example, camellias (*Camellia* spp.) bloom in late fall through winter, and the summer blooms of crapemyrtle in the southeastern United States are notable resources for pollinators (52,53).

Why not breed native plants to be more appealing and marketable? In all plant industries, there are ongoing breeding efforts to improve the esthetics, performance, or growth patterns of plants. Plants of the same species with more unique traits (called cultivars) are commonly patented, trademarked, or branded. Cultivars, and not species, are also what you typically find at a garden center. The same is true for native plants. My residential landscape includes *Itea virginica* (sweetspire) and *Clethra alnifolia* (summersweet), but the only plants available for sale locally were the cultivars, *I. virginica* 'Little Henry' and *C. alnifolia* 'Sixteen Candles.' While these are cultivars, the term "nativar" has entered the plant vernacular to represent cultivars specifically of native plants. In addition to growth and flower differences, variability among cultivars to insects or diseases is also well documented (see Chapter 8). Sometimes, changes in growth habit, disease resistance, and leaf toughness are associated with a reduced response of insects. For example, Japanese beetle feeding damage varies widely between cultivars of the same species (54). Cultivars 'David' and 'Harvest Gold' crabapples are moderately to completely resistant to Japanese beetle feeding, but 'Dolgo' or 'Radiant' can be completely defoliated if planted next to 'David' or 'Harvest Gold.' Similar patterns exist with Japanese beetles and cultivars of crapemyrtle (54). It is a logical extension of plant resistance studies to think that nativars also may differ from the nonselected species.

Dr. Tallamy and his students were among the first to publish research specific to nativars. One study (55) compared 10 species of trees and shrubs with one or more cultivars that varied in leaf color, growth form, fall color, fruiting, or disease resistance. The native plants and their nativar were grown next to one another in a common garden trial and evaluated for feeding damage, insect species diversity, and abundance. Among this group of 10 species, leaf color, leaf variegation, and cultivars developed for disease resistance all changed the abundance or number of insect species. Nativars with pigmented leaves, when the native plant had green leaves, had generally a fewer species and fewer number of insect herbivores. Leaf color and variegation can change the response of plant-feeding insects and mites in ornamental plants (56,57). Selection of plants with unique traits or breeding that reduces insects is not without ecological consequences. Insects on plants are an important resource for higher trophic levels like birds (49). Hundreds of years of plant breeding further suggest that breeding for primarily esthetics will have ecological impacts. Floral forms such as double petals or sterile flowers in *Hydrangea, Prunus,* or *Rose* are not as attractive to bees (51), and larger blooms may be more conspicuous to flower-feeding insects such as Japanese beetles (58).

Talking trees and bad neighbors

Of course, trees cannot use words, but plants communicate using chemicals called volatile organic compounds (VOCs). You may know this phrase from paints or cleaners labeled as "Low VOC." They are chemicals containing carbon that are easily evaporated into the air (volatile). If you like the smell of fresh-cut grass or the rich scents of certain rose blooms, then you enjoy inhaling plant VOC. When these chemicals evaporate from a plant (leaves, flowers, and other plant organs), they become volatiles that may convey information to neighboring plants (59,60) but definitely convey information to insects. This is where the idea of talking trees originated (Box 3.2). If one tree is being fed upon by insects, plant volatiles released by that tree can convey information to neighboring plants resulting in one of two opposite outcomes, plant resistance or susceptibility. Volatiles are great signals because they are noncontact cues that can be carried on the wind away from the damaged plant. Plants do not need to physically touch one another if using volatiles, and insects (herbivores or natural enemies) can detect volatiles and orient toward the plant. If a neighboring plant detects the signal and can respond with enhanced chemical defenses, the outcome would be resistance. Similarly, insects have

BOX 3.2 Talking trees? Really?

Two key papers (59,60) pioneered the idea that a damaged tree could convey an airborne signal (volatiles) to adjacent nondamaged trees. Those nondamaged trees could then increase chemical plant defenses in leaves in the short term (a few days). Those higher levels of chemical defenses in leaves were measurable but neither paper fed the leaves to an insect herbivore to ask if it reduced feeding or if neighboring trees suffered less damage. Before the rise of meta-analysis, Fowler and Lawton (61) searched the published literature at that time and found minimal support for these ideas. Mechanical defoliation was a common substitute for insect feeding in many studies, which negates what we now know as the "spit factor" of insects (62); elicitors in insect salvia that induce volatiles or other changes in plants are not seen with mechanical defoliation. Other studies that use insects for defoliation show some effect on adjacent plants but only in very close proximity (e.g., (63)). Plants are still "talking" using volatiles, but research now focuses on plant-insect communication through volatiles (reviewed in (64)).

elaborate sensors on their antennae for detecting these volatiles in the air. The antennae are the insect nose, and they are moved through air to detect plant volatiles or volatiles from insects like pheromones. If an insect natural enemy responds, the outcome would equally be beneficial (resistance). However, if herbivores (of the same species or a different species) respond, then the outcome would be detrimental (susceptibility). For example, Japanese beetles can respond to plants damaged by other Japanese beetles or unrelated insect herbivores (65).

Plant-produced VOCs are implicated in several of the leading hypotheses explaining host location and distribution patterns, or abundance of insect herbivores and certain natural enemies. Table 3.1 lists the common hypotheses cited to explain patterns of plant-plant and plant-insect interactions. These concepts were developed primarily in studies with field crops or using model plants (*Arabidopsis*) easily manipulated in the lab or greenhouse. However, the ideas seem to translate regardless of the plants or context. The diversity of plants in urban and suburban landscapes make them perfect "laboratories" to test these hypotheses. Regardless of the hypothesis, research strongly supports that a plant growing individually will have a different response from herbivores than plants growing with neighbors (70). Those neighbors can be good or bad for the associated plant. For example, certain host plants may be less likely to be attacked when planted near another plant (associational resistance/good neighbors), whereas the same plant may be more severely attacked when planted adjacent to a different plant (associational susceptibility/bad neighbors). The conclusions across studies with insects and plant neighbors suggest associational susceptibility is more likely than resistance (70).

TABLE 3.1 Key hypotheses or concepts that help explain insect-plant interactions observed in urban landscapes.

Hypothesis or concept	Summary
Resource concentration/host concentration hypothesis (15,66)	This helps to explain why insects tend to be more abundant and problematic in monocultures.
Plant apparency hypothesis (67,68)	• Apparent plants (e.g., woody plants like oak trees) are more likely to be found by insect herbivores. • Unapparent plants (e.g., smaller herbaceous plants) are less likely to be found. • Apparent and unapparent plant categories are based on how human interpret the perception of insects. • Apparent and unapparent plants have different approaches to chemical defenses.
Masking odors hypothesis (69)	Plant odors (volatiles) from a nonhost plant may interfere with perception or location of an associated host plant.
Inappropriate landings hypothesis (34)	• Assumes that nonhost vegetation, and not specifically plant odors, can explain interference in host location. • Herbivores landing on nonhost plants near host plants may cause them to leave the area (patch).
Associational resistance or susceptibility (70,71)	Plants that grow in association with certain other plants can gain resistance to insects, or in some cases may become more susceptible.
Talking trees hypothesis (59–61)	Plants produce volatile organic compounds that provide signals to adjacent plants, insect herbivores, and insect natural enemies.
Plant vigor hypothesis (74)	Fast growing plants or plant parts should produce preferential feeding, greater oviposition, or greater fecundity by insects that use them.
Plant stress hypothesis (75,76)	Plants under abiotic or biotic stress are more suitable for insect herbivores due to the mobilization of nitrogen in leaves or phloem.

These associational interactions are not exclusively explained by volatiles nor is one explanation of these associations applicable to all situations (*34,70*). For example, some plants may physically block other plants making them less visible or provide shade that could reduce the abundance of herbivore or the extent of feeding. Similarly, damage to bluegrass lawns by the hairy chinch bug can be reduced by overseeding with a resistant cultivar of perennial ryegrass (*72*). Red maple trees with crapemyrtle neighbors are 35% defoliated by caterpillars compared with 10% defoliation if planted with no neighbors. In this case, crapemyrtle neighbors somehow caused greater abundance of caterpillars, mainly the greenstriped mapleworm (Fig. 3.5), on associated red maple trees. Other tree neighbors did not cause this dramatic of an increase (*73*).

Hopefully, it is now apparent that the assemblage of plants in urban landscape can change the insect response. This may explain why the same tree may be damaged in the Jones' landscape but rarely attacked in their neighbor's landscape. Unfortunately, the combinations of plants in urban landscapes are endless, and research will never be able to test or understand every situation. In fact, there are many popular magazines, books, and websites dedicated to companion planting in urban landscapes. Despite the many claims, most of those reported plant neighbors have not been tested experimentally and may not stand if rigorously evaluated. Going back to the rose and Japanese beetle study mentioned earlier (*33*), two of the plants in that study were reportedly "good neighbors" for host plants of Japanese beetles. Regardless of what you read online, remember that Japanese beetles, and other plant-feeding insects, are not reading gardening blogs, nor do they subscribe to popular gardening magazines.

FIGURE 3.5 The associated plants in urban landscapes can create unique outcomes depending on the combination. Red maples trees, for example, planted in association with crapemyrtle trees can lead to outbreaks of the greenstriped mapleworm shown here.

Highlights

- Plants are abundant but poor food for insects because most plant parts have low nitrogen content.
- Insect herbivores use nitrogen and secondary plant compounds to determine host range and host quality.
- Plants communicate with each other and with insects mainly through volatile organic compounds.
- Plants growing individually will typically prompt a different response from herbivorous insects than plants growing with neighboring plants.
- Suburban landscapes have high plant diversity, but the diversity creates combinations of plants that could produce associational resistance or susceptibility.

References

1. Zhang, J.; Elser, J. J. Carbon:nitrogen:phosphorus Stoichiometry in Fungi: A Meta-Analysis. *Frontiers in Microbiology* **2017,** *8,* 1281.
2. Mattson, W. J., Jr. Herbivory in Relation to Plant Nitrogen Content. *Annual Review of Ecology and Systemics* **1980,** *11,* 119−161.
3. Tang, Z.; Xu, W.; Zhou, G.; Bai, Y.; Li, J.; Tang, X.; Chen, D.; Liu, Q.; Ma, W.; Xiong, G.; Honglin, H.; He, N.; Guo, Y.; Guo, Q.; Zhu, J.; Han, W.; Hu, H.; Fang, J.; Xie, Z. Patterns of Plant Carbon, Nitrogen, and Phosphorus Concentration in Relation to Productivity in China's Terrestrial Ecosystems. *Proceedings of the National Academy of Sciences, USA* **2018,** *115* (26), 4033−4038.
4. Bernays, E. A. Neural Limitations in Phytophagous Insects: Implications for Diet Breadth and Evolution of Host Affiliation. *Annual Review of Entomology* **2001,** *46,* 703−727.
5. Held, D. W.; Potter, D. A. Characterizing Toxicity of *Pelargonium* Spp. And Two Other Reputedly Toxic Plant Species to Japanese Beetle (Coleoptera: Scarabaeidae). *Environmental Entomology* **2003,** *32* (4), 873−880.
6. Cranshaw, W.; Shetlar, D. *Garden Insects of North America; the Ultimate Guide to Backyard Bugs,* 2nd ed.; Princeton: New Jersey, 2018.
7. Johnson, W. T.; Lyon, H. H. *Insects That Feed on Trees and Shrubs;* Cornell University Press: New York, 1991.
8. Robinson, G.S.; Ackery, P.R. ; Kitching, I.J.; Beccaloni, G.W.; Hernández, L.M., HOSTS - A Database of the World's Lepidopteran Host Plants. Natural History Museum, London. http://www.nhm.ac.uk/hosts.
9. *Every Day Information Source, EDIS;* University of Florida, 2019. http://edis.ifas.ufl.edu.
10. Xu, Y.; Held, D. W.; Hu, X. P. Dietary Choices and Their Implication for Survival and Development of Omnivorous Mole Crickets (Orthoptera: Gryllotalpidae). *Appied Soil Ecology* **2013,** *71,* 65−71.
11. McBride, J.; Jacobs, D. Urban Forest Development: A Case Study, Menlo Park, California. *Urban Ecology* **1976,** *2,* 1−4.
12. Aronson, M. F. J.; Handel, S. N.; La Puma, I. P.; Clemants, S. E. Urbanization Promotes Non-native Woody Plant Species and Diverse Plant Assemblages in the New York Metropolitan Region. *Urban Ecosystems* **2015,** *18,* 31−45.

13. McKinney, M. L. Effects of Urbanization on Species Richness: a Review of Plants and Animals. *Urban Ecosystems* **2008,** *11,* 161–176.

14. Martin, N. A. *A 100% Tree Inventory Using I-Tree Eco Protocol: A Case Study at Auburn University, Alabama.* MS thesis; Auburn University, 2011.

15. Andow, D. A. Vegetational Diversity and Arthropod Population Response. *Annual Review of Entomology* **1991,** *36,* 561–586.

16. Avolio, M. L.; Pataki, D. E.; Gillespie, T. W.; Jenerette, G. D.; McCarthy, H. R.; Pincetl, S.; Clarke, L. W. Tree Diversity in Southern California's Urban Forest: The Interacting Roles of Social and Environmental Variables. *Frontiers in Ecology and Evolution* **2015,** *3,* 73. https://doi.org/10.3389/fevo.2015.00073.

17. Raupp, M. J.; Buckelew, A.; Raupp, E. C. Street Tree Diversity in Eastern North America and It's Potential for Tree Loss to Exotic Borers. *Arboriculture & Urban Forestry* **2006,** *32* (6), 297–304.

18. Sjöman, H.; Östberg, J.; Bühler, O. Diversity and Distribution of the Urban Tree Population in Ten Major Nordic Cities. *Urban Forestry and Urban Greening* **2012,** *11,* 31–39.

19. Cowett, F. D.; Bassuk, N. L. Statewide Assessment of Street Trees in New York State, USA. *Urban Forestry and Urban Greening* **2014,** *13,* 213–220.

20. Milesi, C.; Running, S. W.; Elvidge, C. D.; Dietz, J. B.; Tuttle, B. T.; Nemani, R. R. Mapping and Modeling the Biogeochemical Cycling of Turf Grasses in the United States. *Environmental Management* **2005,** *36* (3), 426–438.

21. Robbins, P.; Birkenholtz, T. Turfgrass Revolution: Measuring the Expansion of the American Lawn. *Land Use Policy* **2003,** *20,* 181–194.

22. Ignatieva, M.; Eriksson, F.; Eriksson, T.; Berg, P.; Hedblom, M. The Lawn as a Social and Cultural Phenomenon in Sweden. *Urban Forestry and Urban Greening* **2017,** *21,* 213–223.

23. Smith, L. S.; Fellowes, M. D. E. The Grass-free Lawn: Management and Species Choice for Optimum Ground Cover and Plant Diversity. *Urban Forestry and Urban Greening* **2014,** *13,* 433–442.

24. Tanner, R. A.; Gange, A. C. Effects of Golf Courses on Local Biodiversity. *Landscape and Urban Planning* **2005,** *71,* 137–146.

25. Yasuda, M.; Koike, F. Do golf Courses Provide a Refuge for Flora and Fauna in Japanese Urban Landscapes? *Landscape and Urban Planning* **2006,** *75,* 58–68.

26. Hodgkison, S.; Hero, J.-M.; Warnken, J. The Efficacy of Small-Scale Conservation Efforts, as Assessed on Australian Golf Courses. *Biological Conservation* **2007,** *136,* 576–586.

27. Larson, J. L.; Redmond, C. T.; Potter, D. A. Impacts of Neonicotinoid, Neonicotinoid-Pyrethroid Premix, and Anthranilic Diamide Insecticides on Four Species of Turf-Inhabiting Beneficial Insects. *Ecotoxicology* **2014,** *23,* 252–259.

28. Kunkel, B. A.; Held, D. W.; Potter, D. A. Impact of Halofenozide, Imidacloprid, and Bendiocarb on Beneficial Invertebrates and Predatory Activity in Turfgrass. *Journal of Economic Entomology* **1999,** *92* (4), 922–930.

29. Gels, J. A.; Held, D. W.; Potter, D. A. Hazards of Insecticides to the Bumble Bees *Bombus impatiens* (Hymenoptera: Apidae) Foraging on Flowering White Clover in Turf. *Journal of Economic Entomology* **2002,** *95,* 722–728.

30. Larson, J. L.; Redmond, C. T.; Potter, D. A. Mowing Mitigates Bioactivity of Neonicotinoid Insecticide in Nectar of Flowering Lawn Weeds and Turfgrass Guttation. *Environmental Toxicology & Chemistry* **2015,** *34* (1), 127–132.

31. Larson, J. L.; Dale, A.; Held, D.; McGraw, B.; Richmond, D. S.; Wickings, K.; Williamson, R. C. Optimizing Pest Management Practices to Conserve Pollinators in Turf Landscapes: Current Practices and Future Research Needs. *Journal of Integrated Pest Management* **2017**, *8* (1), 1−10.

32. Raupp, M. J.; Shrewsbury, P. M.; Holmes, J. J.; Davidson, J. A. Plant Species Diversity and Abundance Affects the Number of Arthropod Pests in Residential Landscapes. *Arboriculture & Urban Forestry* **2001**, *27* (4), 222−229.

33. Held, D. W.; Gonsiska, P.; Potter, D. A. Evaluating Companion Planting and Non-host Masking Odors for Protecting Roses from the Japanese Beetle (Coleoptera: Scarabaeidae). *Journal of Economic Entomology* **2003**, *96* (1), 81−87.

34. Finch, S.; Collier, R. H. Host-plant Selection by Insects-A Theory Based on 'Appropriate/ Inappropriate Landings' by Pest Insects on Cruciferous Plants. *Entomologia Experimentalis et Applicata* **2000**, *96*, 91−102.

35. Raupp, M. J.; Noland, R. M. Implementing Landscape Plant Management Program in Institutional and Residential Settings. *Journal of Arboriculture* **1984**, *10* (6), 161−169.

36. Stewart, C. D.; Braman, S. K.; Sparks, B. L.; Williams-Woodward, J. L.; Wade, G. L.; Latimer, J. G. Comparing an IPM Pilot Program to a Traditional Cover Spray Program in Commercial Landscapes. *Journal of Economic Entomology* **2002**, *95* (4), 789−796.

37. Betancurt, R.; Rovere, A. E.; Ladio, A. H. Incipient Domestication Processed in Multicutural Contexts: A Case Study of Urban Parks in San Carlos de Bariloche (Argentina). *Frontiers in Ecology and Evolution* **2017**, *5,* 166. https://doi.org/10.3389/fevo.2017.00166.

38. Shulaev, V.; Korban, S. S.; Sosinski, B.; Abbott, A. G.; Aldwinckle, H. S.; Folta, K. M.; Iezzoni, A.; Main, D.; Arus, P.; Dandekar, A. M.; Lewers, K.; Brown, S. K.; Davis, T. M.; Gardiner, S. E.; Potter, D.; Veilleux, R. E. Multiple Models for Rosaceae Genomics. *Plant Physiology* **2008**, *147* (3), 985−1003. https://doi.org/10.1104/pp.107.115618.

39. Raupp, M. J.; Davidson, J. A.; Holmes, J. J.; Hellman, J. L. The Concepts of Key Plants in Integrated Pest Management for Landscapes. *Journal of Arboriculture* **1985**, *11* (11), 317−322.

40. Zhang, H.; Jim, C. Y. Contributions of Landscape Trees in Public Housing Estates to Urban Biodiversity in Hong Kong. *Urban Forestry and Urban Greening* **2014**, *13*, 272−284.

41. Coombs, G.; Gilchrist, D., Native and Invasive Plants Sold by the Mid-Atlantic Nursery Industry: A Baseline for Future Comparisons. <https://mtcubacenter.org/action/nurserysurvey/.

42. Pincetl, S.; Prabhu, S. S.; Gillespie, T. W.; Jenerette, G. D.; Pataki, D. E. The Evolution of Tree Nursery Offerings in Los Angeles County over the Last 110 Years. *Landscape and Urban Planning* **2013**, *118,* 10−17.

43. Ballard, M.; Hough-Goldstein, J.; Tallamy, D. Arthropod Communities on Native and Alien Early Successional Plants. *Environmental Entomology* **2013**, *42,* 851−859.

44. Burghardt, K. T.; Tallamy, D. W.; Philips, C.; Shropshire, K. J. Alien Plants Reduce Abundance, Richness, and Host Specialization in Lepidopteran Communities. *Ecosphere* **2010**, *1* (5). https://doi.org/10.1890/ES10-00032.1.

45. Liu, H.; Stiling, P. Testing the Enemy Release Hypothesis: a Review and Meta-Analysis. *Biological Invasions* **2006**, *8,* 1535−1545.

46. Tallamy, D. W.; Shropshire, K. J. Ranking Lepidopteran Use of Native versus Introduced Plants. *Conservation Biology* **2009**, *23* (4), 941−947.

47. Clem, S. C.; Held, D. W. Species Richness of Eruciform Larvae Associated With Native and Alien Plants in the Southeastern United States. *Journal of Insect Conservation* **2015**, *19,* 987−997.

48. Paine, T. D.; Steinbauer, M. J.; Lawson, S. A. Native and Exotic Pests of *Eucalyptus*: A Worldwide Perspective. *Annual Review of Entomology* **2011,** *56,* 181−201.
49. Burghardt, K. T.; Tallamy, D. W.; Shriver, W. G. Impact of Native Plants on Bird and Butterfly Biodiversity in Suburban Landscapes. *Conservation Biology* **2008,** *23,* 219−224.
50. Liebhold, A. M.; Griffin, R. L. The Legacy of Charles Marlatt and Efforts to Limit Plant Pest Invasions. *American Entomologist* **2016,** *62* (4), 218−227.
51. Mach, B. M.; Potter, D. A. Quantifying Bee Assemblages and Attractiveness of Flowering Woody Landscape Plants for Urban Pollinator Conservation. *PLoS One* **2018,** *13* (12), e0208428.
52. Held, D. W.; Chen, Y.; Knox, G.; Pemberton, B.; Carroll, D.; Layton, B. Protecting Pollinators in Urban Areas: Use of Flowering Plants. ANR-2419. *Alabama Cooperative Extension System* **2017,** 8 pg.
53. Riddle, T. C.; Mizell, R. F., III Use of Crape Myrtle, *Lagerstroemia* (Myrtles: Lythraceae), Cultivars as a Pollen Source by Native and Non-native Bees (Hymenoptera: Apidae) in Quincy, Florida. *Florida Entomologist* **2016,** *99* (1), 38−46.
54. Held, D. W. Relative Susceptibility of Woody Landscape Plants to Japanese Beetle (Coleoptera: Scarabaeidae). *Journal of Arboriculture* **2004,** *30* (6), 328−335.
55. Baisden, E. C.; Tallamy, D. W.; Narango, D. L.; Boyle, E. Do Cultivars of Native Plants Support Insect Herbivores? *HortTechnology* **2018,** *28* (5), 596−606.
56. Sadof, C. S.; Raupp, M. J. Effect of Leaf Variegation in *Euonymus japonica* on *Tetranychus urticae* (Acari: Tetranychidae). *Environmental Entomology* **1992,** *21* (4), 827−831.
57. Rowe, W. J.; Potter, D. A.; McNiel, R. E. Suceptiblity of Purple- versus Green-Leaved Cultivars of Woody Landscape Plants to the Japanese Beetle. *HortScience* **2002,** *37* (2), 362−366.
58. Held, D. W.; Potter, D. A. Floral Characteristics Affect Suceptibiliy of Hybrid Tea Roses, *Rosa X Hybrida,* to Japanese Beetles (Coleoptera: Scarabaeidae). *Journal of Economic Entomology* **2004,** *97* (2), 353−360.
59. Rhoades, D. F. Responses of Alder and Willow to Attack by Tent Caterpillars and Webworms: Evidence for Pheromonal Sensitivity of Willows. In *Plant Resistance to Insects;* American Chemical Society, 1983; pp 55−68.
60. Baldwin, I. T.; Schultz, J. C. Rapid Changes in Tree Leaf Chemistry Induced by Damage: Evidence for Communication between Plants. *Science* **1983,** *221* (4607), 277−279.
61. Fowler, S. V.; Lawton, J. H. Rapidly Induced Defenses and Talking Trees: The Devil's Advocate Position. *The American Naturalist* **1985,** *126* (2), 181−195.
62. Alborn, H. T.; Turlings, T. C. J.; Jones, T. H.; Stenhagen, J. H.; Joughrin, J. H.; Tumlinson, J. H. An Elicitor of Plant Volatile From Beet Armyworm Oral Secretion. *Science* **1997,** *276* (5314), 945−949.
63. Dolch, R.; Tscharntke, T. Defoliation of Alders (*Alnus glutinosa*) Affects Herbivory by Leaf Beetles on Undamaged Neighbours. *Oecologia* **2000,** *125,* 504−511.
64. Aljbory, Z.; Chen, M.-S. Indirect Plant Defense Against Insect Herbivores: A Review. *Insect Science* **2018,** *25,* 2−23.
65. Loughrin, J. H.; Potter, D. A.; Hamilton-Kemp, T. R. Volatile Compounds Induced by Herbivory Act as Aggregation Kairomones for the Japanese Beetle (*Popilla japnica* Newman). *Journal of Chemical Ecology* **1995,** *21* (10), 1457−1467.
66. Root, R. B. Organization of a Plant-Arthropod Association in Simple and Diverse Habitats: the Fauna of Collards (*Brassica oleracea*). *Ecological Monographs* **1973,** *43,* 95−124.

67. Feeny, P. Plant Apparency and Chemical Defense. In *Biochemical Interaction Between Plants and Insects;* Wallace, J. W., Mansell, R. L., Eds.; *Recent Advances in Phytochemistry*; Springer: Boston, MA, 1976, Vol. 10.

68. Smilanich, A. M.; Fincher, R. M.; Dyer, L. A. Does Plant Apparency Matter? Thirty Years of Data Provide Limited Support but Reveal Clear Patterns of the Effects of Plant Chemistry on Herbivores. *New Phytologist* **2016,** *210,* 1044−1057.

69. Thiery, D.; Visser, J. H. Masking of Host Plant Odour in the Olfactory Orientation of the Colorado Potato Beetle. *Entomologia Experimentalis et Applicata* **1986,** *41,* 165−172.

70. Barbosa, P.; Hines, J.; Kaplan, I.; Martinson, H.; Szczepaniec, A.; Szendrei, Z. Associational Resistance and Associational Suceptiblity: Having Right or Wrong Neighbors. *Annual Review of Ecology, Evolution and Systematics* **2009,** *40,* 1−20.

71. Tahvanainen, J. O.; Root, R. B. The Influence of Vegetational Diversity on the Population Ecology of a Specialized Herbivore, Phyllotreta Cruciferae (Coleoptera: Chrysomelidae). *Oecologia* **1972,** *10* (4), 321−346.

72. Richmond, D. S.; Niemczyk, H. D.; Shetlar, D. J. Overseeding Endophyitc Perennial Ryegrass into Stands of Kentucky Bluegrass to Manage Bluegrass Billbug (Coleoptera: Curculionidae). *Journal of Economic Entomology* **2000,** *93* (6), 1662−1668.

73. Clem, C. S.; Held, D. W. Associational Interactions between Urban Trees: Are Native Neighbors Better Than Non-natives? *Environmental Entomology* **2018,** *47* (4), 881−889.

74. Price, P. W. The plant vigor hypothesis and herbivore attack. *Oikos* **1991,** *62,* 244−251.

75. White, T. C. R. Plant vigour versus plant stress: a false dichotomy. *Oikos* **2009,** *118,* 807−808.

76. Larsson, S. Stressful time for the plant stress-insect performance hypothesis. *Oikos* **1989,** *56,* 277−283.

Chapter 4

Abiotic factors and interactions with urban pests

We spent much of our time so far considering only the living elements of urban landscapes, but it is now time to consider the roles of nonliving or abiotic elements. These elements include fertilizer, soils, soil moisture, and temperature. This chapter will discuss the importance of these factors and their impact on insects. There may be situations with either suboptimal or above average levels of certain abiotic (heat and water, for example) elements. These conditions influence insects indirectly by altering the vigor or susceptibility of the host plant. Plant stress and plant vigor are poorly understood concepts and will be discussed relative to their known effects on herbivorous insects and mites in urban landscapes.

Temperature

Temperature is one of the most important factors simply because insects are cold-blooded animals. Insect "blood" called hemolymph sloshes around inside their body. Except for veins in the wings and one central artery, insects have no veins for circulation. Hemolymph is also generally clear since it does not carry oxygen like mammalian blood. Cold blooded does not refer to the temperature of hemolymph. Instead, you get an insect's body temperature by ramming a sharp probe into the thorax or abdomen and recording the body temperature, but not specifically the blood. The insect body temperature is similar to air temperature when at rest. And when they are on leaves in the sun, their body temperature is generally higher than the air temperature. This is why it is easier to catch a dragonfly or beetle when they are sitting in the shade or on a leaf on a cool morning in the fall. Now that you have a brief introduction to insect energetics, let us talk about the consequences of being cold blooded for insect development, survival, and behavior.

The development rate (time to develop) of immature insects will decrease as temperatures increase to a point where temperature stops development or becomes lethal. Similarly, development time will increase with a decrease in temperature to a point where development stops or the insect dies. With insects, development time decreases incrementally less with increased temperature and will never be zero because each developmental stage requires a

Urban Landscape Entomology. https://doi.org/10.1016/B978-0-12-813071-1.00004-X

minimum amount of time to complete. We have temperature-dependent development data on several important pests in urban landscapes, including mole crickets, Japanese beetles, other introduced scarab species, imported fire ants, many thrips and aphid species, most common caterpillars, and azalea lace bugs. With azalea lace bug development (Fig. 4.1), the lowest development time is about 20–25 days between 27 and 30°C (80–86°F). Most eggs will not hatch at a constant temperature of 33°C (91°F) or more, or at temperatures below 15°C (59°F). Females do not produce eggs below 15°C (59°F) (*1*). Temperature can also limit adult survival. Female and male adult lace bugs survive >120 days, on average, at 20.6°C (70°F) and have the lowest average survival of 20–30 days at 31.7°C (89°F) (*2*). These experiments exclude the effects of natural enemies which should have a greater impact on insects at temperatures that prolong that particular life stage. This is called the slow-growth high mortality hypothesis (*3*). If a natural enemy prefers, or can only use one or two life stages of an insect for oviposition or development, then temperatures that prolong that life stage enhance the activity of the natural enemy. Furthermore, if temperatures increase and the pest develops more quickly through those stages, then the natural enemy will be less effective. There are small parasitoid wasps (*Anagrus* sp.) that attack the eggs (and only eggs) of azalea lace bugs. In spring, if temperatures rise from 21 to 24°C (70–75°F), the lace bugs will develop 10 days faster through the egg stage (*1*), fewer eggs will be available, and the wasps will be less effective. That is how temperature may influence development and the effectiveness of natural enemies through the application of the slow-growth mortality hypothesis. Scientific reviews (*4*) and other experimental studies (*5,6*) suggest

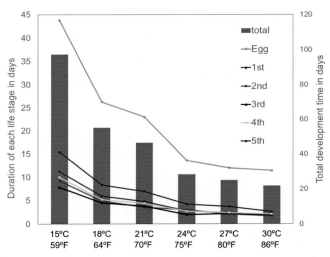

FIGURE 4.1   Development time of azalea lace bugs on azalea relative to temperature (*1*).

that the slow-growth-high-mortality hypothesis may not apply to all situations. In particular, insect herbivores that have one generation per year (univoltine), herbivores that live underground, or parasitoids that can only develop in a narrow range of insect host species may be less likely to adhere to the principles of slow-growth high mortality.

Temperature is also variable and subject to localized effects of impervious surfaces especially within the urban core areas of cities. These effects are often referred to as heat islands. Geosciences and biologists have found common ground in the study of these urban phenomena. These heat island effects are also not just a phenomenon of large metropolitan areas. Downtown Auburn, a midsized city in east Alabama and a large retail area in Opelika, AL are both heat islands (7). If there is a shopping mall or large retail complex in an otherwise rural area, then you may have a heat island, and local heating. One of the best-studied effects of heat islands on urban pests was conducted on populations of gloomy scale, an armored scale on red maple trees, in urban areas around Raleigh, NC (5). Gloomy scale is a univoltine insect that is a specialist on maple trees. In this study, Dale and Frank (5) looked at the relative effects of urban temperatures and biotic factors (natural enemies and vegetation). At warmer temperatures, female scales are larger in size and produce more eggs. In this situation, temperature is the driving force behind large outbreaks of this scale insect in urban heat islands. This stands as a counterpoint to the long-accepted idea that natural enemies are one of the most important forces limiting or regulating herbivore populations (8,9). This study on urban insects and heat islands provides new ideas on how urban settings may affect insect populations. In a follow-up paper, Dale and Frank (9) reviewed the literature and identified the limits in our understanding of urban warming in urban ecosystems.

Insects can also behave differently at different temperatures usually to adjust their body temperatures. Since the body temperature of insects is dependent on the environment, insects can behave in ways to thermoregulate their temperature. While this information is well documented for several insects like honey bees, we will consider examples of pest species common in urban landscapes in North America. Imported fire ants (*Solenopsis invicta* and *Solenopsis richteri*) are introduced pests that produce mounds in turfgrass and ornamental plantings. Mounds (Fig. 4.2) are most apparent and largest in spring and fall when soil temperatures are coolest. They are generally less apparent in summer when soil temperatures are highest except during overcast and rainy periods (10). The mounds function to increase heat relative to surrounding soil like a solar collector. Mounded soil warms 15% faster than surrounding soil and are constructed to maintain a "hot spot" near the center as the sun passes overhead (11). Red imported fire ants (*S. invicta*) have preferred temperatures for rearing immatures (brood) between 30 and 32°C (86—90°F), and foragers/workers prefer temperatures slightly cooler which helps to increase their life span (12). Workers move within soil and the mounds to adjust temperature and

FIGURE 4.2   Imported fire ants make surface mounds that act like solar collectors enabling the colony to maintain temperatures suitable for development.

to avoid temperatures exceeding 32°C (90°F) at which brood would die (*12*). It is also speculated that workers likely move brood east to west inside surface mounds to track patterns of solar warming of the mound surface (*11*). Japanese beetle adults move in and out of the shade on host plants during the day and may bask in the sun in the morning to warm their bodies enough to fly. When daytime temperatures are <20°C (68°F), beetles will fall to the ground and not be able to fly (*13*). Beetles that land on the ground without the ability to fly are more susceptible to ground active predators like ants and spiders (*14*). This beetle is commonly referenced as "sun loving" because it appears to also defoliate plants from the top down (*15*). This pattern of plant feeding may be explained because a beetle is trying to position itself in the sun to warm more quickly in the morning to be able to fly. There are only a few examples of thermoregulation behavior for urban landscape insects, mostly ants, but these patterns may prove important to understanding management. Some Extension publications, for example, suggest tapping Japanese beetle adults into a bucket of soapy water in the morning. However, this will not work in most parts of the southernmost range of this pest because nighttime temperatures rarely drop below 20°C (68°F) during the beetle flight. Finally, thermoregulation is not a biological switch that flips, there are other factors, particularly feeding, that can also influence these behaviors. For example, southern chinch bugs commonly aggregate and do damage near paved sidewalks in the spring. This pattern may be due to temperature regulation, but it may be because the grasses closest to the heat of the sidewalk break dormancy faster, and are the first food plants to conduct sap on which these insects feed.

Soils

Soils are a composite of living and nonliving components with unique chemical and biological interactions. Entire books are written on the dynamics of soil, but I only want to include the necessary terms and background needed to help explain the impacts on insects. Soil particles are classified by size into stone, sand, silt, and clay. Stone and sand are the largest particles and clay is the smallest. Relative to one another, think of sand as a beach ball compared to clay which would be a dime on the same scale. The relative amount of those particles can create various soil textures like the loamy sandy soils in the southeastern United States. There are other components like organic matter, plants, microbes (fungi and bacteria), insects and mites, and nematodes. And then there are the chemical reactions driven in part by changes in pH, an indication of the acidic or basic nature of soils. Soil pH is the main piece of information home gardeners wish to learn when they submit a soil test to the local Extension office for analysis.

Insects use soil primarily for development, analogous to the development in water by immature stages of insects like mosquitoes. The influence of soils on urban landscape insects are best documented for turfgrass insects like ants, white grubs, and mole crickets. Soil interactions with insects are best illustrated as a two-way street where soil factors influence abundance and survival of insects, and soil insects influence attributes of soil. This section will exclude both drought, covered in a separate section later in this chapter, and a discussion of earthworms. Earthworms are well-studied invertebrates in soil, but this section will focus on soil arthropods. We will consider the influence of soil excavation, compaction, texture, moisture, microbes, and pH on soil insects. Soil excavation is obvious with insects like ants and mole crickets that construct subterranean tunnels. Mole crickets excavate soil below ground and leave deposits of soil on the surface (Fig. 4.3). In 7 days, a single adult southern mole cricket will excavate 126.5 g (4.4 oz) of clay or 141 g (8.3 oz) of loamy sand while constructing an underground tunnel network. Mole crickets can make longer and more branched tunnels in loamy sand soils compared to tunnels in clay soil. This is why we can see activity of mole

FIGURE 4.3   The activity of mole crickets such as making surface mounds shown here is directly related to soil moisture.

crickets north of the coast but usually just in locations where the soil profile is amended with sand. Mole crickets individually create a branched tunnel network about ≥60 cm (24 in.) long and 2−3 times their body width (*16*). Ants, with thousands of workers, excavate larger soil amounts (*17*). Tunneling by mole crickets and ants can also change how water moves through soil. The subterranean tunnels create a soil profile that roughly resembles Swiss cheese. This allows water to flow rapidly from the surface through those holes (*16,18*) causing the surface layer of soil to dry faster.

Mounding behavior of ants in soils changes soil moisture and nutritional content. And, ants are widespread in urban landscapes worldwide. Even a casual observer could note the conspicuous ring of darker green grass that can surround the mounds of ants like fire ants (Fig. 4.4). This results from changes induced by mounds on adjacent soils. A review of 106 studies on ants noted no effects of ant mounds on soil pH, but root and overall plant growth were greater relative to adjacent soil with no ant mounds (*18*). As with any review, the results of individual studies may disagree or agree with the trends found in these reviews. For example, red imported fire ants cause increased plant growth in nest soil, with soil acidification shown in some studies (*19*), but not in others (*20*).

Soil moisture is a critical component for egg laying, development, and behavior of soil insects. Soil moisture thresholds exist for egg laying and hatch, but can vary by soil texture class (*21*). Eggs of white grubs absorb soil water, and a minimum soil moisture of 3%−9% is needed for hatch of masked chafer or Japanese beetle eggs (*21,22*). White grub species vary in tolerance to soil moisture, with Japanese beetle grubs being less tolerant of lower soil moisture than masked chafers (*23*). Mole crickets can also survive in soils with low soil moisture for several weeks, but the tunnel structure is not stable at soil

FIGURE 4.4 Soil-dwelling insects can also affect soil properties and fertility for plants. The green halos around these imported fire ant mounds show how ants can change soil nutrition and improve plant growth adjacent to mounds.

moistures $\leq 2\%$ (24). For perspective, soil moistures of 5% or less are rare if the grass is green and vigorous. The immature stages and especially the initial stages of egg development are the most sensitive to extremely dry soils. Egg-laying females of scarab beetles and mole crickets in turfgrass are more discriminating of soil moisture. Female mole crickets lay eggs in response to increasing soil moisture and seem to opportunistically respond to short-term bursts in soil moisture (24). Female masked chafers do not lay eggs in dry soils ($\leq 15\%$ soil moisture) (22). Oviposition preference based on soil moisture selection by females is likely tied to greater sensitivity of eggs to desiccation or enhanced growth of immature stages in soils that have consistently more moisture content (22,23).

The behavior of insects in soil is also influenced by soil properties. Mole crickets move up and down within their subterranean tunnels over time. As they tunnel, they will avoid tunneling through compacted soil and can detect and avoid pathogenic fungi in the upper layers of the soil profile (25). The surface activity of mole crickets is also correlated with soil moisture (26). This is why some insecticides labeled for use against mole crickets suggest irrigating turfgrass the night before the application. This should encourage more surface activity and the possibility that mole crickets will contact the insecticide and be killed. Similar to mole crickets, white grubs can also move up and down in the soil profile. They naturally do this in response to seasonal changes, going deeper in soil for the winter and moving up in the soil in spring in response to temperatures (27). White grubs also move in response to short-term changes in soil moisture and temperature (28). Furthermore, different species of grubs will "roam" variable distances within the soil profile depending on larval stage or in response to changes in soil conditions or disturbances (28,29).

Plant fertilization and insect responses

All plants, as well as insects, have variable carbon, nitrogen, and phosphorus content. These are the building blocks for life on Earth. Carbon is harvested from the atmosphere as CO_2 by plants and converted to sugars and biomass through photosynthesis for use by consumers like insects. Leaves are mostly carbon with relatively little nitrogen. Similarly, plants pump water from soils into the atmosphere through transpiration. Water is equally critical to insects, and plants have 50%−70% water content in leaves. Water and nitrogen content are usually greater in new flushes of plant growth (30). Since insects consume a mixture of elements and water, it may be easier to think of plant hosts for herbivores as having a carbon to nitrogen ratio (C:N). The C:N ratios of leaves from common landscape trees range 20−50:1 (31,32), and 10−30:1 in grass foliage (33) (Fig. 4.5). These ranges reflect variation between species but also the influence that local growing conditions (soil nutrients or soil type) can have on the C:N ratio for the same plant species (31). Insects consuming leaves

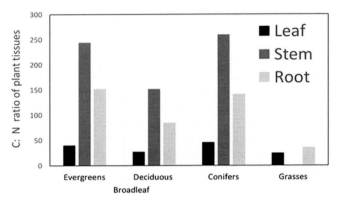

FIGURE 4.5   Insects need nitrogen from plants but are mostly consuming carbon. The type of plant and plant parts have very different carbon to nitrogen ratios (C:N) (*31–33*).

have the problem of consuming excess carbon or sometimes water while trying to access nitrogen. However, leaves are also about the best plant part to use as food if the goal is to minimize surplus carbon intake (Fig. 4.5). Stem and root feeders on broadleaf plants have a greater challenge to access nitrogen, but this is less of an ecological issue for insects consuming grass roots and leaves. Nitrogen is present in all the amino acids needed for life which means you cannot have protein, RNA, or DNA without nitrogen. Insect herbivores are adapted to the low nitrogen content of their hosts. A study compared the food plants and frass (insect poop) of 130 caterpillar species representing 15 common families. Most of the host plants were trees (oak, willow, pine) with a few perennials, and two grasses. The C:N ratio of caterpillar frass strongly reflects the C:N ratio of the plant they consume (*34*). Sap-feeding insects like aphids produce a liquid frass called honeydew that enables them to excrete excess water as well as extra carbon as sugars.

What does nitrogen added to plants do to insect herbivores? Nitrogen content of plant leaves is increased in the short term through supplemental fertilization. Fertilization is a common practice in urban landscapes and with horticultural plants and turfgrasses. As a result of fertilization, free amino acids in leaves will increase (*35*). Insect feeding amount (damage), abundance, or performance (growth, longevity, or fecundity) are responses used to gauge the influence of fertilization on host quality. Let us begin by looking at the trends or predictions found in scientific reviews then focus specifically on studies with urban landscape plants and insects.

Predictions about supplemental nitrogen and responses of insects

1. *Nitrogen applied alone or as a fertilizer mixed with phosphorus and potassium has a positive effect on insects.* The positive effects on herbivores are increased abundance, development, life span, or greater numbers of offspring (*35−38*). Only a low proportion of studies ever find negative effects of nitrogen fertilization on insects and mites.

2. *Insects with sucking mouthparts may have a greater positive response to nitrogen fertilization (36,37,39).* Sucking insects are commonly associated with "flush feeding." Flush feeders are insect herbivores that appear to prefer fast-growing, new shoots (Fig. 4.6). Flushing feeding is predicted by the Plant Vigor Hypothesis (*40*). Stems and leaves on vigorously growing, fertilized plants are most likely to be attacked (*39*). This preference explains why experienced landscape managers know first to check new shoots on trees and shrubs for pests like aphids.

3. *Insect taxonomy may predict responses to nitrogen fertilization.* Insects in the orders Hemiptera (sucking insects) and Lepidoptera (caterpillars) should have the greatest positive benefit from the addition of nitrogen to their host plants. Diptera (leafminers or gall makers) and Coleoptera (beetles) should also have positive responses to nitrogen fertilization but not as strong as the other orders. Plant-feeding species in Orthoptera (grasshoppers, katydids) are predicted to have a negative response to nitrogen fertilizer (*36*). Within our discussion of orders and mouthparts is the concept of feeding guilds, insects across orders that have similar life histories. This is a way to summarize insects that use the same plant habitat (leaf rollers, leaf chewers, borers, etc). In addition to sapsucking and leaf chewing, guilds including leafminers and gall inducers should benefit from nitrogen fertilization but not stem borers, flower feeders, or leaf folders (*39*).

FIGURE 4.6 Aphids are considered one example of a "flush feeder." They exploit the fast-growing plant parts which typically are of better quality.

How do these predictions compare to studies on pests found in urban landscapes?

There are several studies positively linking nitrogen fertilizer to positive responses by fall armyworm (Lepidoptera) and chinch bugs (Hemiptera: *Blissus* spp.) in turfgrass. Turfgrass fertilizer rates are commonly expressed in lb of nitrogen (N) per 1000 sqft. Lower rates (0.5—1 lb N per 1000 sqft) have limited effects but high rates (≥ 2 lbs n per 1000 sqft) can significantly increase populations of southern chinch bugs in fertilized St. Augustine grass (*41*) and Western chinch bug in buffalograss (*42*). Fall armyworm larval survival increases by 60%—97% and larval weight increases by as much as 76% in fertilized bermudagrass compared to nonfertilized bermudagrass (*43*). There are limited data on other turf-infesting species, except for white grubs. Japanese beetles and masked chafer grubs weigh the same, were at the same developmental stage, and similarly abundant in nonfertilized and nitrogen fertilized tall fescue (*23*). These few studies with turfgrass insects support the general trends noted in the broader literature.

The data on the fertilization effects with woody and herbaceous plants are not as simple as in turfgrass. The published experiments evaluate a wide range of fertilizer treatments (constant feeding, slow release, liquids, or granular) depending on the ornamental plant, and most are conducted in greenhouses using potted plants. This makes it more difficult to generalize about pests or even plants because each study uses a "recipe" specific to the study. Furthermore, very few studies applied fertilizer to existing plantings. Butler et al. (*36*) suggested that experimental results in greenhouses may be less likely to find significant results with nitrogen fertilization than those under natural conditions. It does not mean those studies are not valid, but that greenhouse data may not accurately represent data from field studies. Let us consider the variability in the few published studies with two-spotted spider mites and fertilizer applications to ornamental plants. Constant feeding roses with 150 ppm N will produce 1.1% greater concentration of foliar nitrogen and greater plant growth in those roses, but it also virtually doubles the number of mites and eggs noted on roses constant fed with 50 or 75 ppm N (*44*). However, nitrogen fertilization sufficient to produce a high quality potted ivy geranium did not positively affect two-spotted mite populations (*45*). Burning bushes (*Euonymus alata*) can be locally very susceptible to two-spotted spider mites. Infested and fertilized burning bushes did not have more two-spotted mites than nonfertilized plants (*46*).

Potted maples fertilized with 20 or 40 g N per tree per year produce more damage from potato leafhopper (*Empoasca fabae*) and greater abundance of maple spider mites (*Oligonychus aceris*) compared to nonfertilized trees. However, trees that received the 40 g N rate had nearly three times the damage from potato leafhopper as nonfertilized trees (*47*). Azalea lace bugs (*Stephanitis pyrioides*) prefer plants with higher nitrogen but feeding on nitrogen fertilized plants does not increase injury nor does it provide other advantages (faster development time or increased survival) (*48,49*). High fertilization rates of woody plants in the nursery may temporarily increase pest populations or damage in the first few months after planting. A study with "Sutyzam" crabapple trees applied high rates of N fertilizer to potted trees than evaluated growth or caterpillars feeding on those trees before and after planting. Larval weight gain increased with fertilizer application rate even for larvae feeding on the leaves of those trees 3 months after planting (*50*). This is the only published study that followed high input plants from production into the landscape.

I commonly ask students in class if they think a fertilized plant will have more or less damage and why? On one hand, it is a better quality host and may support longer feeder or greater populations. On the other hand, there may be less damage to fertilized plants because insects are better at converting plant mass to body mass at higher nitrogen contents of plant tissue (*30*). A herbivore should consume less plant sap or tissue of a fertilized plant because it has more nitrogen per "bite" and they can more efficiently convert it to body mass. Despite this, there is a perception that nitrogen fertilization improves plant vitality and in turn, positively improves resistance to diseases and insects (*51*). I see this entrenched paradigm particularly in the marketing of fertilizers to the general public. However, the literature provides little to no support for this paradigm among woody plants. In fact, there is support for decreased resistance among fertilized woody plants (*38*). There is equally no support that fertilization increased tolerance to defoliation (*38*). Nitrogen fertilization in woody plants and grasses can affect certain groups of secondary plant compounds that provide resistance to certain insects. In woody plants, supplemental nitrogen generally decreases foliar concentrations of the secondary plant compounds which could render plants more susceptible to plant-feeding insects (*35*). Based on the limited number of studies with pests of turfgrass and ornamentals, the effects of fertilizer seem to follow the same trends noted in reviews for a wide range of crops (*35–38*).

What about other sources of plant nutrients?

It seems that every few years there is either a new research paper on silicon supplements or a product being marketed that contain silicon. These products can alleviate biotic (plant diseases) and abiotic (drought) stresses in certain plants (*52*). Silicon naturally occurs in soils but is not always bioavailable. Sand for example is a form of silicon but it is not bioavailable in that form. Therefore, some soils may contain silicon but still require silicon supplements. Silicon in soil solution is taken up in the transpiration stream and then moves via xylem to either roots or shoots. Once inside the plant parts, it remains and does not move (*52*). The goal of adding silicon is to increase foliar leaf concentrations such that biotic or abiotic stress is reduced. The molecular and biomechanical effects from silicon fertilization of plants can produce leaves or roots with increased "toughness." Plant "toughness" (sclerophylly) can alter the feeding behavior of plant-feeding insects like caterpillars (*53*) or beetles (*54*). Plant defense theory defines these as elemental defenses. Certain plants can hyperaccumulate elements like nickel or silicon, and use them for their defense. In addition to plant toughness, metal-accumulating plants can also be toxic to insects that consume them.

The question remains, *"Can we manipulate these as nutritional/elemental defenses in plants?"* First, all plants cannot accumulate silicon. Silicon accumulation can vary by plant genus but also by species within the same genus. Si accumulation is more common among grasses. Tall fescue (*Festuca arundinacea*), for example, is a silicon-accumulating species with variation in uptake among varieties (*55*). The application of silicon to grasses or ornamental plants that are not silicon accumulators may explain why the results of published experiments are so variable. Soil applications of silicon also provide the highest foliar silicon concentrations compared to foliar applications (*56*). Silicon application to bermudagrass, centipede grass, seashore paspalum, zoysia, St. Augustine grass, and creeping bentgrass can significantly increase foliar silicon content but those increases fail to have negative impacts on tropical sod webworm (*Herpetogramma* spp.), black cutworm, or masked chafer grubs (*57,58*). Increased silica content can cause mandibular wear in the mouthparts of turf-feeding caterpillars (*58,59*). None of the tests with silicon added to turfgrass produced mortality or other negative impacts on pests; however, application of sodium silicate to corn yielded greater mortality of fall armyworms fed leaves from Si-fertilized foliage compared to nontreated corn (*59*). There are a few published studies with ornamental plants that are equally variable. Silicon fertilization of ornamental coleus failed to have negative consequences for populations of citrus mealybug (*Planococcus citri*) reared on treated plants (*60*). Silicon fertilizer applied to hybrid sunflower significantly increases foliar content, yet that change was not sufficient to cause mortality or prolong development of *Chlosyne lacinia saundersii,* a key caterpillar pest of sunflower in Brazil. Larvae reared on silicon-treated sunflower leaves had significantly reduced weight after 10 and 15 days (*61*). A constant feeding (every 2 days) of potted ornamental zinnia with potassium silicate increased foliar silicon, but did not reduce survivorship of green peach aphids. Over 20 days, aphids produce fewer offspring on silicon-treated plants. While this could reduce the rate that aphid populations grow, plants regularly fertilized with silicon did not prevent feeding or kill green peach aphids directly (*62*). As noted for grasses, ornamental plants that are not reportedly Si-accumulating species should not be expected to increase silicon content of foliage or have negative consequences for developing insects with chewing or sucking mouthparts. The application to roots and the need to constantly feed (*62*) silicon to plants to achieve possible effects may limit, or make cost-prohibitive, the adoption of silicon fertilization for pest management.

Organic amendments including organic fertilizers, humic and fulvic (organic) acids, bacteria, and bacterial metabolites are other groups with reported effects on insects. Organic fertilizers in turfgrass have been studied relative to white grubs and certain foliage-feeding pests. Most of our common white grub species are root feeders and incidentally consume organic matter and soil as they feed. However, some species of grubs (Green June beetles and black turfgrass ataenius) more readily fed on organic matter. Female Green June beetles (*Cotinis nitida*) and the grubs orient toward organic matter in the soil but seem particularly attracted to broiler litter from chickens, cow manure, and hay (*63*). Organic fertilizers like stabilized sewage sediment or meat meal products were not attractive to female black turfgrass ataenius (*Ataenius spretulus*) but can result in significantly greater abundance of grubs of that species where they are applied (*64*). However, the same stabilized sewage sediment product applied to turfgrass can increase populations of southern chinch bugs (*41*). Biochar is another soil amendment being marketed to arborists and landscape managers. Biochar is not charcoal, but it is a heat-treated organic product that appears similar to charcoal. Most of the published literature indicates benefits of biochar on root growth and enhanced growth of beneficial soil microbes and microbial communities. And in some cases, biochar is used as a carrier for beneficial microbes (*65*). The biochar literature is far better developed relative to disease suppression and root growth promotion than it is for insects. Insect studies exist with food crops but only one preliminary laboratory study with ornamentals or turfgrass (*66*). In that study, different predatory and wood-boring beetles and ants in a Petri dish were confined with biochar for 48 h. Some insects that come in direct contact with biochar in lab conditions may be negatively affected but not through indirect effects on soil quality or tree growth. Unfortunately, the current scientific literature has very little support for or against biochar, and the impact on urban landscape pests.

My research program has some of the only data on the effects of soil microbes on turfgrass and turfgrass pests. This work is with plant growth–promoting rhizobacteria (root bacteria). These microbes are mainly bacilli bacterial (e.g., *Bacillus, Paenibacillus*) that can be cultured and placed in water for application to grasses and ornamentals. These microbes are not the nitrogen-fixing *Rhizobium* widely used on seeds of annual crops. Although we lump them into this large group called rhizobacteria, there are many different species and even strains of species. A strain is essentially a bacterial cultivar. Strains of the same bacterial species can behave differently in the root zone. Therefore, research with specific strains cannot be applied broadly to other strains of that same species or even more broadly to all rhizobacteria. Therefore, you need to review the data on beneficial rhizobacteria carefully because of the specific nature of strains and blends. Most of the work in my program is with Blend 20, a mixture of *Bacillus pumilus* AP7, *Bacillus pumilus* AP18, and *Bacillus sphaericus* AP282. The bacterial strains, indicated as AP numbers, in our work are beneficial microbes that colonize roots and sometime aboveground parts of plants resulting in a unique, temporary plant phenotype. If you are familiar with the word endophyte, some of these bacteria create endophytic phenotypes of grasses. This unique phenotype may have enhanced growth and impacts on insects. The rhizobacteria we have evaluated promote aboveground and root growth in bermudagrass (67,68) and tall fescue (69). In bermudagrass, the bacteria in Blend 20, DH32, and DH44 enter the grass within 24—72 h and can be found above and below ground for weeks after one application (70). The most obvious effect of growth promotion is increased root growth (67). This can produce a grass that has enhanced tolerance to root-feeding insects like mole crickets and white grubs (69,71) and enhanced tolerance to plant-parasitic nematodes. Above ground, we also have noted changes in the response of fall armyworm females. Moths lay fewer eggs on grass treated with Blend 20 and our other rhizobacteria strains (68).

Plant stresses and insect responses

"Stress" as a blanket term is grossly nonspecific, highly situational, and requires knowledge of the right place for the right plant. Florida dogwood, for example, is intolerant of flooding but not red maple (72). Florida dogwood is typically an understory tree, yet often planted in full sun exposure in urban landscapes which can increase attacks by dogwood borers (73). Similarly, hosta lilies, an herbaceous perennial, grow best in shade, but shade can reduce growth and carbon-based defenses in woody plants (35). Plant stress can have abiotic or biotic causes, and be temporal or seasonal. Flooding, damage from diseases or insects, or air pollutants could happen one month or year, but not at another time. Stresses, like temperature and drought, light exposure and temperature, are confounding with other stresses, and not easily separated without complex statistical analyses (5). These reasons likely explain why I commonly hear plant stress as a scapegoat for plant disfunction among landscape and turfgrass managers. For many problems, "stress" is the cause without fully understanding if the plant is stressed or if the pest in question even responds positively to stress. Do not worry, you are in good company. Many professors at your state university do not have a good answer either, unless that specific pest, or plant-stress-pest combination, have been well studied. The next section will outline the principles of plant stress through the eyes of plant ecologists, and hopefully refine our understanding of plant stress and its effects on insect abundance in urban landscapes.

Plant stress is often defined by the agent causing the stress. The literature outlines abiotic (water stress, temperature stress, fertility stress, pollutant stress) and biotic stresses (competition, damage from insects or disease). The Plant Stress Hypothesis (74) outlines the principles for outbreaks of insects on plants under stress. In one of the first reviews. White (74) used specific examples to support the main idea that stressed tissues are different from normal plant tissues. The hypothesis suggests that stress should prompt leaves to act like they do in the fall, when they mobilize nitrogen as free amino acids and move it away from those leaves. When nitrogen is mobilized in plants, it should be more readily available to herbivores that opportunistically exploit stressed plants. Since White (74), other research has both supported and refuted parts of this hypothesis. As mentioned before, the Plant Stress Hypothesis appears to resonate with professional plant people attempting to explain insect outbreaks. There are thousands of research papers on plant stress and insects in virtually every possible plant production system. Many of those papers investigate crops, forest trees are the next most studied plant system, and ornamental plants and turfgrass are the least studied in the literature. Furthermore, even fewer water stress studies with woody plants or ornamentals have a field component. It is easier to manipulate water consistently in a pot, and so studies on established plants or in-ground plants are a small minority among the already small number of stress studies with ornamentals and turfgrass. Plant size is another concern with direct application of studies conducted with whips or seedlings to more mature plants in urban landscapes. Would a 1 ft (0.3 m) tall tree or shrub in a pot represent what is happening with a mature tree in the landscape? Not likely. For example, sap-sucking insects benefit more by attacking stressed saplings compared to mature trees, and colonization of wood borers is more successful in mature trees than saplings (75). As we did in the fertilization section, let us begin with some of the predictions made across all the published research using review papers, then we will compare those to some selected research specifically with urban landscape pests and plants.

Expected plant and herbivore responses to stress

1. *Plant stresses vary and so do the responses of plants and insects*

This was an early criticism of the Plant Stress Hypothesis and one that persists today. White (74) listed many different types of stress, including water stress, fertility stress, and pollution as examples that supported his hypothesis. Later reviews (76) noted that it was difficult to generalize across many different stresses. Most review articles seek to identify overall trends in plant stress and herbivore response, but often they also separate individual stress responses when reported. While water stress is the most widely studied, there are also considerable research devoted to pollution, light, and biotic stresses.

2. *Stress changes plant nutritional value for arthropod herbivores*

As student, I could never understand why an insect would want to colonize a plant in decline. In my mind this was like insects jumping onto a sinking ship. However, Huberty and Denno (77) suggested a possible mechanism for how drought and recovery could be beneficial to insects or mites on stressed plants. Water stress appears to change plants more than shading or ozone exposure (*35*). Waring and Cobb (*37*) first suggested that natural drought (often chronic or longer term) provided different outcomes than short-term drought. Short-term drought is sometimes listed as intermittent or cyclic. Both drought types occur in natural and produce different results in plants. Chronic drought is generally not beneficial for either plants or herbivores, with the exception of cambium feeders (discussed in point #4). Under water stress, plant cells begin to lose turgor pressure (e.g., lower leaf water content and wilting) and eventually it reaches some threshold (likely unique for each genus or species) where the drought impedes the ability of sapsuckers to feed (*77*). Water stress and recovery (cyclic) presents more opportunities for sapsuckers to benefit from mobilized nitrogen and sugar in plant cells (*77*). In support of this mechanism, insect feeding coincident with stress can increase their reproduction. However, sap-sucking insects would have decreased reproduction if a host plant is released from the stress before they begin feeding (*75*).

Fig. 4.7 is a visualization of what may happen in a water-stressed plant. This shows 3 weeks of imposed water stress (weeks 1–3) and 3 weeks of poststress recovery (weeks 4–6) on Yukon hybrid bermudagrass in a greenhouse experiment (*78*). Electrolyte leakage approximates when nitrogen, amino acids, and sugars are more mobile or available to insects and mites. The horizontal dashed line is a fictitious threshold of leaf water content for a sapsucking arthropod like bermudagrass mites or chinch bugs. Below that line, there is not enough turgor pressure (water content) to allow feeding. The gray boxes indicate times during water stress and recovery when cells may have higher than normal levels of nitrogen or sugars. Populations of bermudagrass mites may exist on this grass. When water stress occurs, there is a period of

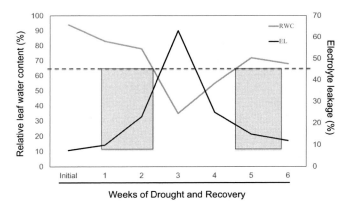

FIGURE 4.7 Relative water content (RWC) and electrolyte leakage (EL) in hybrid bermudagrass in a greenhouse experiment (*78*) subjected to 3 weeks of imposed water stress (weeks 1–3) and 3 weeks of poststress recovery (weeks 4–6).

time when populations could dramatically increase while nitrogen and sugars are enhanced in the plant. If that cycle runs over 12 or 16 weeks, there are multiple opportunities for a plant feeder to have short-term population booms. Cyclic drought and recovery is likely and more common in urban landscapes than crops because of the prevalence of irrigation systems. Some horticulturalists and arborists preach watering landscape plants deeply and infrequently. Extending what we just learned about cyclic drought, infrequent watering could create unintended consequences for urban landscape pests. Infrequent watering could create a drought and recovery cycle that promotes periods of mobilized nitrogen and sugars. If we stretch that cycle out over time, it could create boom and bust cycles for some pests. This is a consideration and not a call for a *coup d'état* against this long-standing horticultural principle. There simply are limited data to support or refute that this is occurring but the outcome is strongly predicted. The few studies on this topic with urban landscape plants are discussed later in the water stress section.

3. *Plant growth rate and type provides insight into the responses by insects*

There is evidence for reduced survival, growth, fecundity for chewing insects on plants that are slower growing relative to those that are faster growing (75). Growth rate is assumed to represent different groups of plants, trees as slower growing, and herbaceous plants as faster growing. There are more stress experiments with insects using stressed trees or woody plants, and fewer with herbaceous ornamental plants. Across all studies, plant ecology also influences insect responses to plant stress. Conifers and deciduous plants respond differently to stress. Conifers, more so than deciduous trees and shrubs, are more likely to have compromised health and in turn greater susceptibility to pests or more plant damage (79). Drought is expected to change the susceptibility of conifers to benefit herbivores particularly sapsuckers (37,75). Conversely, drought-induced changes to shrubs and herbaceous plants would be less beneficial to insects (37).

4. *Wood borers (cambium feeders) are expected to benefit from using stressed plants*

First suggested by Larsson (80), this has since been supported by experimental studies and meta-analyses (75). Drought is a strong contributor to borer colonization of angiosperms trees (37), but it is not the only trigger.

5. *Sap-sucking insects are expected to benefit from using stressed hosts*

Larsson (80) also predicted that sapsucking insects would benefit from using host plants under stress. Among the various stresses, pollutants can potentially increase the offspring of sap-sucking insects (75). Despite that, most of the published research with sap-sucking insects investigates water stress. Studies indicate both positive and negative effects of water stress on sap-sucking insects. The type of plant sap being used may also influence these results (77). Phloem feeders (e.g., aphids) are expected to be highly responsive to the cyclic water stress previously discussed.

6. *Stress should negatively affect gall inducers*

Gall-inducing or gall-forming insects are a guild that represents six orders of insects and mites. As a guild, they must exploit actively growing plant parts to highjack the new growth for their development. Price (40) specifically discussed how gall-inducing insects benefit from plant vigor. Among the reviews, most include negative trends or significant negative survival and density for gall formers on stressed plants (75,77).

7. *Leaf-chewing insects and leafminers should be least affected by plant stress*

Chewing insects generally have overall nonsignificant effects when all guilds (leafminers, leaf chewers) are lumped together (75,77). Leafmining insects are specialized chewing insects that develop inside a plant. Effects of plant stress are either positive (75) or inconclusive (77) for leafminers. Staley et al. (81) subjected five species of leafminers to the same water regimes and each responded differently. Free-living leaf chewers (caterpillar and beetles) are expected to avoid stressed plants and would have negative reproduction and survival effects if they consume stressed plants (75,77).

Pollutants

The most well-documented interaction of urban air quality and insects is the study of industrial melanism among peppered moths (*Biston betularia*) in industrialized England in the late 1800s and early 1900s (82,83). Populations of this peppered moth have multiple color forms including white and white speckled forms, and a black (melanic) form. Melanic simply refers to the darkened wings or other parts of the insect through pigmentation (82). This melanic form was not caused by pollution, in fact, hundreds of species of moths naturally have this type of color variation. In response to the air pollution, there were fewer lichens and more soot on the urban trees. The color variants that were white were more conspicuous and, over time in the polluted environment, became less dominant in the population. Cook and Saccheri (83) provide a good summary of the ecological and genetic mechanisms that contributed to the shift to the melanic form.

Pollutants in the air and soil can affect insects, but much of the work seems focused on air pollutants (e.g., fluorides, sulfur, sulfur and carbon dioxide, ozone, and dusts) and aboveground insects (84). Furthermore, these studies are rarely conducted in urban environments. Many investigate insect population changes in laboratory studies or on sites with high environmental impacts such as mining locations (79). Pollutants can affect both herbivores and natural enemies either directly or indirectly (affecting the insect host) (79,84,85).

Plants compromised by pollutants can be more susceptible to insects, or natural enemies may be negatively impacted from consuming insects tainted, but not killed, by pollutants (79,85). Leaf-chewing and sap-sucking insects tend to respond with greater populations in polluted sites compared to nonpolluted sites. Also, the density of gall-inducing insects is typically greater on polluted sites (79). Pollution may also enhance the reproductive success of sap-sucking insects when they are directly exposed to the pollutant but decreased performance of leaf-chewing insects directly exposed to pollutants (75). Studies on insect parasitoids suggest either no effect or negative effects, with a few suggesting parasitoids or parasitism are enhanced by pollutants. Ozone effects are largely negative for parasitoids (85). Dusts are interesting because they do not necessarily have a toxic component. Dusts or airborne sediments serve as desiccants to smaller (parasitoids) and soft-bodied insects. Dusts can locally increase the abundance of certain arthropods (mites and insects) along gravel roads, for example (84). Finally, if you recall VOC are the ways that plants "call" to insects and for insects to chemically communicate with one another. These airborne signals can be weakened or degraded in the presence of air pollutants in urban environments (86). These are mainly laboratory studies, but they suggest air pollution may be an underestimated driver of reduced parasitism in the inner city relative to rural or suburban sites.

Light

Light intensity and patterns of shading can influence plant-feeding insects and their natural enemies. Certain secondary plant compounds like flavonoids accumulate in plant tissue under high light conditions. These primarily function like plant sunscreen, protecting the genetic material of the plant cell from UV light. However, these plant compounds may also provide protection against plant-feeding insects (87). High amounts of flavonoids in zoysiagrass are associated with resistance to fall armyworms (87). Although grasses are typically grown in full sun, resistant cultivars of zoysiagrass in shade may have reduced amounts of flavonoids and could lose resistance to fall armyworms. Light may also activate certain secondary plant compounds. For this reason, certain caterpillars feeding on umbelliferous plants (anise, fennel, and their relatives) tend to roll the leaves to limit exposure to light (88). Shading of woody plants can also effectively reduce carbon-based foliar secondary plant compounds (35,89). This should make sense because trees in shade will be less photosynthetically active and then have less carbon to make the carbon-based defenses. Light-activated compounds, flavonoids and coumarins, are reduced in woody plants growing in shade (35).

Plants in full sun can harvest more carbon dioxide enabling them to make more carbon-based plant defenses and sugars. Sugar is the first chemical stopping point of photosynthesis, and therefore leaves in the sun should have more sugars than leaves in the shade. This was experimentally tested with roses and Japanese beetles (*90*). When presented out of context in the laboratory, the beetle consumed a greater amount of rose leaves from plants in the sun versus those placed under shade. Leaves from roses in the full sun have higher sugar content than leaves from roses grown under shade (*90*). Sugars are universal feeding stimulants for plant-feeding insects. In another experiment (*89*), potted white oak (*Quercus alba*) seedlings grow under full sun or full shade from bud break to fully expanded leaves were used to determine effects on the development of caterpillars. Both types of potted plants (sun and shade) were then placed into sunny or shaded locations in the forest understory. Caterpillars that consume leaves from trees grown in the sun perform better, but caterpillars feeding in the sun also suffered greater attacks by parasitoids than caterpillars feeding in the shade (*89*). Light intensity is also a significant factor in the susceptibility of St. Augustine grass to southern chinch bugs. Lawns exposed to a high light intensity (about 950 lumens per sqft) will have more chinch bugs than mostly shaded lawns (*91*).

Water stress

Statistically, there is always at least one region within the United States at some level of drought (https://droughtmonitor.unl.edu). Furthermore, moderate droughts have occurred annually in the United States since 2000, and about 6.5% of the land in the United States has been in extreme drought since 2000 (*92*). While much of the water stress research focuses on a reduction in water, we need to begin with those pests that benefit in well or excess water situations (usually high rainfall). Ambrosia beetles (Scolytinae: *Xylosandrus* and *Cnestus* spp.) are responsive to ethanol produced by waterlogged trees that are intolerant to temporal flooding (*93*). When there is above average rainfall in spring and summers, the abundance of sharpshooters and spittlebugs on turfgrass and trees is apparent. These are sap-sucking insects that use xylem (vascular tissue that conducts mostly water and minerals) and this dynamic is noted anecdotally for several species (*94–97*). Rainfall for soil moisture is critical to the activity and survival of many soil insects. Crane flies, turfgrass pests primarily in the northeastern and northwestern United States, must have adequate rainfall in September and October or larval populations will suffer (*98*). Surface activity of mole crickets, another common soil-dwelling pest, is positively correlated with increased soil moisture (*26*). During temporal droughts, the surface activity of mole crickets is limited and restricted to irrigated areas that are often overwatered (*personal observations*). Finally damage from white grubs can be masked by high soil moisture. Grub densities necessary to damage cool season grass are lower in nonirrigated turfgrass than irrigated turfgrass (*99*).

The following studies represent a diversity of experimental methods and observations of natural responses of mites, thrips, aphids, lace bugs, leaf chewers, wood borers, to drought (*100*) or supplemental irrigation (e.g.,

Refs. *101–104*) in woody plants and turfgrass. Six selected studies (Table 4.1) investigated the responses of two spider mites (*Tetranychus urticae*), clover mites (*Bryobia* sp.), and honeylocust spider mite (*Platytetranychus multi-digitali*) to natural and imposed drought. Only Mason et al. (*46*) evaluated cyclic versus chronic water stress but mites were similar in both water stress treatments.

Six selected studies (Table 4.2) show the responses of sap-sucking aphids and armored scale insects (Hemiptera: Sternorrhyncha) to natural and imposed drought on woody ornamentals. One interesting note is the potential interaction between resistance to aphids and drought. Ramirez and Verdugo (*105*) evaluated hybrid poplars that varied in levels of resistance to aphids, then subjected them to varying water treatments. Under drought conditions tolerance was enhanced but resistance was compromised in the hybrids known for

TABLE 4.1 Select studies where responses of mites were reported under imposed or natural water stress.

Arthropod	Plant	Brief synopsis of methods	Key result(s)
Two-spotted spider mites (*101*)	Burning bush	Potted plants provided variable irrigation	Populations greater and increase more rapidly at lowest irrigation levels
Two-spotted spider mites (46)	Burning bush	Pot-in-pot plants provided variable fertilizer and water	No difference in fertilizer levels. Well-watered plants had more infested leaves than water stressed
Two-spotted spider mites (*102*)	Butterfly bush	Potted plants provided variable irrigation	No population difference but more damage on water-stressed plants
Honeylocust spider mite (*103*)	Honeylocust	Potted trees provided variable irrigation	No effect of water stress
Spider mites (*100*)		Observations of populations under natural drought	Greater populations
Clover mites (*104*)	Ky bluegrass	Field plots provided variable irrigation	Plots not provided supplemental irrigation had twice as many mites as irrigated plots

TABLE 4.2 Select studies where responses of aphids and armored scale insects were reported under imposed or natural water stress.

	Plant	Brief synopsis of methods	Key result(s)
Aphids			
	Hybrid poplar	Small trees that vary in resistance in pots were exposed to aphids and variable irrigation (105)	Tolerance to aphids was greater under water stress but resistance was reduced
	Northern red oak	Established landscape trees under natural moisture stress (106)	Population increase but low overall counts
		Observations of populations under natural drought (100)	Greater populations
Armored scales			
Gloomy scales	Maple	Manipulated temperature and drought of trees in landscapes (5)	Temperature (main effect) increases size and number of offspring and effects enhanced by drought
White peach scale	Mulberry	Trees growing in urban environments (107)	Survival negatively affected by drought
Euonymus scale	Euonymus	Greenhouse test with rooted cuttings (108)	Mortality (or fecundity) was lower on water-stressed plants

resistance. There will be an expanded discussion about host plant resistance in Chapter 8 but the interaction of drought and plant resistance to insects was worth noting here.

Seven selected studies (Table 4.3) show the responses of thrips and true bugs (Hemiptera: Heteroptera) to natural and imposed drought on ornamentals and turfgrass. Thrips were more abundant on plants under water stress. Despite the importance of lace bugs in urban landscapes in the United States, relatively little information is reported on their responses to drought in their hosts. Chrysanthemum lace bugs and oak lace bugs are related (*Corythucha* spp.). Adults of *Corythucha* spp. may be able to detect and avoid water-stressed leaves; however, there may be survival and developmental advantages to using water-stressed plants.

TABLE 4.3 Select studies where responses of thrips and true bugs (Heteroptera) were reported under imposed or natural water stress.

Thrips		Observations of populations under natural drought (100)	Greater populations
Thrips	Ky bluegrass	Field plots provided variable irrigation (104)	Thrips were less abundant in irrigated plots
Western flower thrips	Chrysanthemum	Potted plants provided variable irrigation and fertilizer in greenhouse (109)	Fewer adult thrips in highest irrigation. Fewer adults and immatures in lowest fertilizer treatment
True bugs (Heteroptera)			
Oak lace bug	White oak	Potted sapling subject to water stress (110)	Adults avoided leaves from water-stressed plants
Oak lace bug	Northern red oak	Established landscape trees under natural moisture stress (106)	Population increase but low overall counts
Chrysanthemum lace bug	Goldenrod	Potted plants maintained under variable water levels (111)	Insects did not prefer water-stressed to normal plants. Greater survival of nymphs to adult stage and greater adult body mass on water-stressed plants
Southern chinch bug	St. Augustine grass	Potted grass provided variable irrigation and levels of insects (112)	No interaction between irrigation and chinch bug density; damaged regardless of irrigation treatment
Hairy chinch bugs		Observations of populations under natural drought (100)	Greater populations

There is a strong consensus relative to drought and the benefits to the guild of wood borers (79,113). Much of our knowledge of borers of woody landscape plants is largely based on the extensive literature on trunk and shoot borers in managed forests. The forest entomology literature has well-established links between water stress and borer attacks with wood borers performing better on water-stressed trees (*114*). Although, at least one paper has questioned the direct relationships between drought and tree mortality from destructive species like southern pine beetle (*115*). Other information on borers in urban landscapes comes from the influx of new research following the introduction of exotic species like emerald ash borer, Asian long-horned beetles, or granulate ambrosia beetles.

Wood-boring insects in urban landscapes are most commonly moths or beetles that are borers as larvae. However, the literature and our understanding of borer–stress interactions in woody plants is based mostly on research with beetles. Furthermore, despite the diversity of wood-boring beetles worldwide, only a few genera (*Ips, Dendroctonus, Agrilus, and Phoracantha*) are well studied. The selected studies (Table 4.4) are much of the available literature on borer–water stress interactions of urban woody plants. There are other studies where trees were damaged to induce stress or water stress, but they will be discussed separately. Two common concepts emerge from the research on wood-boring beetles: (1) adult females can exploit volatiles produced by damaged or drought-stressed trees (*121*), and (2) drought enables or improves colonization of the tree or growth of the developing larva inside the tree (*119,122*). Trunk and branch borers that are caterpillars are represented mainly by two families: Sesiidae (clearwing moths) and Cossidae (zebra and carpenterworm moths). There are too few studies with drought stress and wood-boring caterpillars to draw broad conclusions. Moths may exploit trunk wounds (*73,125*) for egg laying. However, evidence that water stress alters host location by the adult moths similar to wood-boring beetles is observational or anecdotal (*126,127*). The diverse numbers of insects casually lumped together as "wood borers" use different host location strategies. We should be careful when extrapolating information from wood-boring beetles to wood-boring caterpillars. Another group of moth borers (tip-boring moths) exploit the new growth of trees and shrubs. The success of these insects would be expected on vigorous plants and not stressed plant with reduced growth.

Five selected studies (Table 4.5) show the responses of leaf-consuming insects to naturally occurring and imposed drought on ornamentals and grasses. As expected, most of the studies use caterpillars, with one study on leafcutter ants. For reference, leafcutter ants do not consume the leaves they harvest. The trends across these studies suggest either no effect or that caterpillar species benefit from using host plants under water stress.

These studies represent virtually all of the currently published research on water stress with ornamental plants and turfgrass when coupled with insect and mite pests. As stated before, the diversity of plant material and pests in landscapes creates reliance on studies with forest trees and crops to supplement our knowledge of how groups of pests (e.g., borers, sapsuckers) respond to water and other stresses in urban environments. Hopefully, it is now clear why I suggested earlier that plant care professionals should not overuse "plant stress" to explain all pest outbreaks.

TABLE 4.4 Select studies where responses of wood-boring beetles and moths to woody plants under imposed or natural water stress.

Long-horned beetles (cerambycidae)

Eucalyptus long-horned borer	Eucalyptus (gum trees)	8-yr-old trees in field provided three water treatments (116)	Water stress reduces bark water content, but increased sugar content and larval weight
Eucalyptus long-horned borer	Eucalyptus (gum trees)	Potted trees subject to drought treatment (117)	Short-term drought rendered trees susceptible but watering can restore resistance
Eucalyptus long-horned borer	Twelve Eucalyptus spp.	Evaluated planted trees that vary in drought tolerance for insect resistance (118)	Species differ in susceptibility; attacks are relative to water stress
Eucalyptus long-horned borer	Eucalyptus spp.	Evaluated bark moisture and success of larvae in living trees and logs of two species (119)	Borer larvae could only colonize logs and water-stressed trees (low moisture content in bark)
Old house borer (Hylotrupes bajulus)	Scots pine	4-yr-old trees in the field subjected to two levels of water stress (120)	Drought did not affect the growth rate
Asian long-horned beetle	Acer spp. (maples)	Review article on the biology of this borer (121)	Adults attracted to VOC from stressed trees, particularly boxelder

Metallic wood-boring beetles (Buprestidae)

Emerald ash borer	Black ash, Manchurian ash	Potted trees exposed to water stress (122)	Low water availability increases larval mass
Bronze birch borer	Betula spp. (birch)	Review article on the biology of this borer (123)	Documented outbreaks associated with widespread loss of birch
Goldspotted oak borer	Coast live oak	Compared natural levels of drought stress to tree attacks (124)	Adult does not select water-stressed trees but borer can cause water stress
Moths, lilac/ash borer	Green ash	Established landscape trees under natural moisture stress (106)	Increase incidence of borers on trees under moisture stress

TABLE 4.5 Select studies where responses of leaf-consuming insects to woody and herbaceous plants under imposed or natural water stress.

Monarch larvae	Milkweed	Greenhouse and field experiments (128)	Larvae in the field grew larger on plants with reduced rainfall
Fall armyworm		Observations of populations and natural drought (129)	High populations anecdotally linked to drought, especially rainy periods after drought
African cotton leafworm	Apple	Variable water stress; preference and performance tests with insects (130)	Greater acceptance of leaves and greater growth rate with high water stress
Leafcutter ants	Cake bush (Piper marginatum)	Potted plants and a laboratory colony of leafcutter ants (131)	Ants harvest twice as much leaf area from water-stressed versus well-watered plants
Tussock moth, eastern tent caterpillar, and gypsy moth	Crabapple	Potted trees placed under variable and fertilizer treatments then transplanted (50)	No effect of irrigation on weight gain
Gypsy moth and tussock moth	Black poplar	Potted, rooted cuttings exposed to variable fertility and water (132)	Drought reduced growth of gypsy moth larvae

Stress induced by damage

Plant damage from previous insect feeding, disease, or even a careless person with a string trimmer can also affect plant susceptibility to insects (73,125). In these studies, researchers typically induced defoliation or a trunk wound, and then measured plant growth and insect responses. Infestation of Florida dogwood trees in urban landscapes can be as high as 60%, mainly due to trunk injuries from lawn maintenance equipment (125). A series of papers have investigated the response of metallic wood-boring beetles (Buprestidae) to girdled trees. Girdling in landscape trees varies in depth and severity but the wounds interfere with movement of water and plant metabolites. A tree can be girdled in a way to cut only xylem or phloem, or both vascular tissues. Girdling smaller caliper trees would be expected to be more devastating than on large trees. Girdling trees increases attractiveness to two-lined chestnut borer and emerald ash borer (133,134). Xylem girdling of oak trees produces a greater response from two-lined chestnut borer than phloem girdling.

Interestingly, two-lined chestnut borer respond in as little as 4 h after a host tree is girdled, and girdled nonhost trees do not attract two-lined chestnut borer (*133*). The process of harvesting trees in a nursery may also change the attractiveness of trees. Root pruning as well as digging then replanting can increase the attractiveness of maple trees to flatheaded apple tree borers (*135*). There are recurring discussions among academics and plant professionals about damage from herbicides, resulting plants stress, and attacks by borers. Herbicides are used by landscape managers and in production nurseries. There is only one published study (*134*) with emerald ash borer on this topic. Direct sprays of ash tree trunks with a nonselective herbicide (triclopyr) in diesel fuel can induce stress that recruits emerald ash borers to those trees.

Damage to landscape plants may also occur from previous attack by other insects or infection by plant diseases. Federal entomologists in the early 1900s observed that virus-infected plants and peach trees, trees infected with Dutch elm disease, and plants that had a prior infestation appeared more attractive to, or had greater feeding damage from, Japanese beetles (*27*). However, follow-up experiments suggest these interactions are subtler. Japanese beetles can cue in on plants that have recently been damage by Japanese beetles (*136*) or tree-feeding caterpillars (*137*). Feeding by herbivores promotes the release of VOC that are attractive to adult Japanese beetles. But these biological cues are time sensitive. New feeding damage is not as attractive as damage from the previous day (*137*), and leaves that reflush after complete defoliation from eastern tent caterpillar sustain similar damage to trees that were not defoliated (*138*). Defoliation is another way to experimentally stress trees for research. For example, 80% defoliation of smaller caliper red maples can significantly recruit more flatheaded apple tree borers than trees without defoliation (*135*).

Landscape plants may have multiple insect, or insect and disease complexes, yet these are rarely studied at the same time. Mostly entomologists evaluate plants for insect attack and plant pathologists may do similar, yet independent evaluations for diseases. This area of research is a gap in our understanding of plant pests in urban landscapes. For example, does boxwood leafminer infestation affect boxwood blight and vice versa? Urban stresses are equally important for plant diseases like *Hypoxylon* canker (*139*). One interdisciplinary study investigated *Thyronectria* canker—infected honeylocust trees and attack in a 2-year experiment. In this example, there was no effect on responses by foliage-feeding caterpillars or metallic wood-boring beetles to diseased trees (*140*). However, these interactions may occur over several seasons or cycles of stress. This is very evident to anyone who has watched trees die over a series of years from either drought, disease, or insects. This slow decline may actually be a series of decline and recovery steps. The Tree Decline Recovery Seesaw model (*141*) was developed as a model for tree farm managers to apply in managing cycles of abiotic and biotic stresses. But, it also provides a useful framework on which more holistic research considering multiple urban stresses could be built.

Highlights

- Temperature influences the development rate of insects, behaviors, and success in landscapes.
- Temperature and moisture content are important for insects that use soil to develop such as white grubs and mole crickets.
- Supplemental nitrogen to plants will mostly favor the development and survival of insects.
- Plant stress from pollution, water, drought, or damage is overused as the cause for insect outbreaks.

References

1. Braman, S. K.; Pendley, A. F.; Sparks, B.; Hudson, W. G. Thermal Requirements for Development, Population Trends, and Parasitism of Azalea Lace Bug (Heteroptera: Tingidae). *Journal of Economic Entomology* **1992,** *85* (3), 870–877.
2. Neal, J. W., Jr.; Douglass, L. W. Development, Oviposition Rate, Longevity, and Voltinism of *Stephanitis pyrioides* (Heteroptera: Tingidae), an Adventive Pest of Azalea at Three Temperatures. *Environmental Entomology* **1988,** *17,* 827–831.
3. Feeny, P. Plant Apparency and Chemical Defense. In *Biochemical Interaction Between Plants and Insects;* Wallace, J. W., Mansell, R. L., Eds.; *Recent Advances in Phytochemistry*; Springer: Boston, MA, 1976, Vol. 10.
4. Chen, K.-W.; Chen, Y. Slow-growth High-Mortality: A Meta-Analysis for Insects. *Insect Science* **2018,** *25,* 337–351.
5. Dale, A. G.; Frank, S. D. The Effects of Urban Warming on Herbivore Abundance and Street Tree Condition. *PLoS One* **2014,** *9* (7), e102996. https://doi.org/10.1371/journal.pone.0102996.
6. Hourston, J. E.; Bennett, A. E.; Johnson, S. N.; Gange, A. C. Does the Slow-Growth, High-Mortality Hypothesis Apply below Ground? *PLoS One* **2016,** *11* (8), e0161904 https://doi.org/10.1371/journal.pone.0161904.
7. Hug, A. W. *The Study of Urban Heat Islands in the Birmingham and Auburn-Opelika, Alabama Urban Areas, Using Satellite and Observational Techniques.* MS thesis; Auburn University, 2014.
8. Russell, E. P. Enemies Hypothesis: A Review of the Effect of Vegetational Diversity on Predatory Insects and Parasitoids. *Environmental Entomology* **1989,** *18* (4), 590–599.
9. Dale, A. G.; Frank, S. D. Urban Plants and Climate Drive Unique Arthropod Interactions and Unpredictable Consequences. *Current Opinion in Insect Science* **2018,** *29,* 27–33.
10. Vogt, J. T.; Reed, J. T.; Brown, R. L. Temporal Foraging Activity of Selected Ant Species in Northern Mississippi during Summer Months. *Journal of Entomological Science* **2004,** *39* (3), 444–452.
11. Vogt, J. T.; Wallet, B.; Coy, S. Dynamic Thermal Structure of Imported Fire Ant Mounds. *Journal of Insect Science* **2008,** *8,* 31.
12. Porter, S. D.; Tschinkel, W. R. Fire Ant Thermal Preferences: Behavioral Control of Growth and Metabolism. *Behavioral Ecology and Sociobiology* **1993,** *32* (5), 321–329.
13. Kreuger, B.; Potter, D. A. Diel Feeding Activity and Thermoregulation by Japanese Beetles (Coleoptera: Scarabaeidae) within Host Plant Canopies. *Environmental Entomology* **2001,** *30,* 172–180.

14. Held, D. W.; Potter, D. A. Characterizing Toxicity of *Pelargonium* Spp. And Two Other Reputedly Toxic Plant Species to Japanese Beetle (Coleoptera: Scarabaeidae). *Environmental Entomology* **2003**, *32* (4), 873–880.
15. Rowe, W. J.; Potter, D. A. Vertical Stratification of Feeding by Japanese Beetles within Linden Tree Canopies: Selective Foraging or Height Per Se? *Oecologia* **1996**, *108*, 459–466.
16. Bailey, D. L.; Held, D. W.; Kalra, A.; Twarakavi, N.; Arriaga, F. Biopores from Mole Crickets (*Scapteriscus* spp.) Increase Soil Hydraulic Conductivity and Infiltration Rates. *Applied Soil Ecology* **2015**, *94*, 7–14.
17. Frouz, J.; Jilková, V. The Effects of Ants on Soil Properties and Processes (Hymenoptera: Formicidae). *Myrmecological News* **2008**, *11*, 191–199.
18. Farji-Brener, A. G.; Werenkraut, V. The Effects of Ant Nests on Soil Fertility and Plant Performance. *Journal of Animal Ecology* **2017**, *86*, 866–877.
19. Lafleur, B.; Hooper-Bui, L. M.; Mumma, P. E.; Geaghan, J. P. Soil Fertility and Plant Growth in Soils from Pine Forests and Plantations: Effect of Invasive Red Imported Fire Ants *Solenopsis invicta*. *Pedobiologia* **2005**, *49*, 415–423.
20. Lockaby, B. G.; Adams, J. C. Pedoturbation of a Forest Soil by Fire Ants. *Soil Science Society of America Journal* **1985**, *49*, 220–223.
21. Régnière, J.; Rabb, R. L.; Stinner, R. E. *Popillia japonica*: the Effect of Soil Moisture and Texture on Survival and Development of Eggs and First Instar Grubs. *Environmental Entomology* **1981**, *10*, 654–660.
22. Potter, D. A.; Gordon, F. C. Susceptibility of *Cyclocephala immaculata* (Coleoptera: Scarabaeidae) Eggs and Immatures to Heat and Drought in Turf Grass. *Environmental Entomology* **1984**, *13*, 794–799.
23. Potter, D. A.; Powell, A. J.; Spicer, P. G.; Williams, D. W. Cultural Practices Affect Root-Feeding White Grubs (Coleoptera: Scarabaeidae) in Turfgrass. *Journal of Economic Entomology* **1996**, *89*, 156–164.
24. Hertl, P. T.; Brandenburg, R. L.; Babercheck, M. E. Effect of Soil Moisture on Ovipositional Behavior in the Southern Mole Cricket (Orthoptera: Gryllotalpidae). *Environmental Entomology* **2001**, *30* (3), 466–473.
25. Villani, M. G.; Allee, L. L.; Preston-Wilsey, L.; Consolie, N.; Xia, Y.; Brandenburg, R. L. Use of Radiography and Tunnel Castings for Observing Mole Cricket (Orthoptera: Gryllotalpidae) Behavior in Soil. *American Entomologist* **2002**, *48*, 42–50.
26. Hertl, P. T.; Brandenburg, R. L. Effect of Soil Moisture and Time of Year on Mole Cricket (Orthoptera: Gryllotalpidae) Surface Tunneling. *Environmental Entomology* **2002**, *31*, 476–481.
27. Fleming, W. E. *Biology of the Japanese Beetle, Tech. Bull. No. 1449;* US Department of Agriculture: Washington, 1972.
28. Villani, M. G.; Wright, R. J. Use of Radiography in Behavioral Studies of Turfgrass-Infesting Scarab Grub Species (Coleoptera: Scarabaeidae). *Bulletin of the Entomological Society of America* **1988**, *34* (3), 132–144.
29. Villani, M. G.; Nyrop, J. P. Age-dependent Movement Patterns of Japanese Beetle and European Chafer (Coleoptera: Scarabaeidae) Grubs in Soil-Turfgrass Microcosms. *Environmental Entomology* **1991**, *20* (1), 241–251.
30. Schoonhoven, L. M.; van Loon, J. J. A.; Dicke, M. *Insect-plant Biology;* Oxford University Press: New York, 2005.

31. Hu, B.; Simon, J.; Günthardt-Goerg, M. S.; Arend, M.; Kuster, T. M.; Rennenberg, H. Changes in the Dynamics of Foliar N Metabolites in Oak Saplings by Drought and Air Warming Depend on Species and Soil Type. *PLoS One* **2015,** *10* (5), e0126701.

32. Tang, Z.; Xu, W.; Zhou, G.; Bai, Y.; Li, J.; Tang, X.; Chen, D.; Liu, Q.; Ma, W.; Xiong, G.; Honglin, H.; He, N.; Guo, Y.; Guo, Q.; Zhu, J.; Han, W.; Hu, H.; Fang, J.; Xie, Z. Patterns of Plant Carbon, Nitrogen, and Phosphorus Concentration in Relation to Productivity in China's Terrestrial Ecosystems. *Proceedings of the National Academy of Sciences, USA* **2018,** *115* (26), 4033–4038.

33. Hamido, S. A.; Guertal, E. A.; Wood, C. W. Seasonal Variation of Carbon and Nitrogen Emissions from Turfgrass. *American Journal of Climate Change* **2016,** *5,* 448–463.

34. Kagata, H.; Ohgushi, T. Carbon to Nitrogen Excretion Ratio in Lepidopteran Larvae: Relative Importance of Ecological Stoichiometry and Metabolic Scaling. *Oikos* **2012,** *121,* 1869–1877.

35. Koricheva, J.; Larsson, S.; Haukioja, E.; Keinänen, M. Regulation of Woody Plant Secondary Metabolism by Resource Availability: Hypothesis Testing by Means of Meta-Analysis. *Oikos* **1998,** *83,* 212–226.

36. Butler, J.; Garratt, M. P. D.; Leather, S. R. Fertilisers and Insect Herbivores: a Meta-Analysis. *Annals of Applied Biology* **2012,** *161,* 223–233.

37. Waring, G. L.; Cobb, N. S. The Impact of Plant Stress on Herbivore Population Dynamics. In *Insect–plant Interactions,* Vol. 4, Bernays, E., Ed.; CRC Press: Boca Raton, USA, 1992; pp 167–226.

38. Herms, D. A. Effects of Fertilization on Insect Resistance of Woody Ornamental Plants: Reassessing an Entrenched Paradigm. *Environmental Entomology* **2002,** *31* (6), 923–933.

39. Cornelissen, T.; Fernandes, G. W.; Vasconcellos-Neto, J. Size Does Matter: Variation in Herbivory between and within Plants and the Plant Vigor Hypothesis. *Oikos* **2008,** *117* (8), 1121–1130.

40. Price, P. W. The Plant Vigor Hypothesis and Herbivore Attack. *Oikos* **1991,** *62,* 244–251.

41. Busey, P.; Snyder, G. H. Population Outbreak of the Southern Chinch Bug Is Regulated by Fertilization. *International Turfgrass Society Research Journal* **1993,** *7,* 353–361.

42. Carstens, J.; Heng-Moss, T.; Baxendale, F.; Gaussoin, R.; Frank, K.; Young, L. Influence of Buffalograss Management Practices on Western Chinch Bug and its Beneficial Arthropods. *Journal of Economic Entomology* **2007,** *100,* 136–147.

43. Lynch, R. E. Effect of Coastal Bermudagrass Fertilization Levels and Age of Regrowth on Fall Armyworm (Lepidoptera: Noctuidae) Larval Biology and Adult Fecundity. *Journal of Economic Entomology* **1984,** *77,* 948–953.

44. Chow, A.; Chau, A.; Heinz, K. M. Reducing Fertilization for Cut Roses: Effect on Crop Productivity and Twospotted Spider Mite Abundance, Distribution, and Management. *Journal of Economic Entomology* **2009,** *102,* 1896–1907.

45. Chen, Y.; Opit, G. P.; Jonas, V. M.; Williams, K. A.; Nechols, J. R.; Margolies, D. C. Twospotted Spider Mite Population Level, Distribution, and Damage on Ivy geranium in Response to Different Nitrogen and Phosphorus Fertilization Regimes. *Journal of Economic Entomology* **2007,** *100,* 1821–1830.

46. Mason, N. R.; Potter, D.; McNiel, R. What Factors Affect Twospotted Spider Mite Populations on Burning Bush? *Proceedings of the Southern Nursery Association Research Conference* **1998,** *43,* 179–182.

47. Prado, J.; Quesada, C.; Gosney, M.; Mickelbart, M. V.; Sadof, C. Effects of Nitrogen Fertilization on Potato Leafhopper (Hemiptera: Cicadellidae) and Maple Spider Mite

(Acari: Tetranychidae) on Nursery-Grown Maples. *Journal of Economic Entomology* **2015,** *108* (3), 1221—1227.

48. Casey, C. A.; Raupp, M. J. Effect of Supplemental Nitrogen Fertilization on the Movement and Injury of Azalea Lace Bug (*Stephanitis pyrioides* (Scott)) to Container-Grown Azaleas. *Journal of Environmental Horticulture* **1999,** *17* (2), 95—98.

49. Casey, C. A.; Raupp, M. J. Supplemental Nitrogen Fertilization of Containerized Azalea Does Not Affect Performance of Azalea Lace Bug (Hemiptera: Tingidae). *Environmental Entomology* **1999,** *28* (6), 998—1003.

50. Lloyd, J. E.; Herms, D. A.; Rose, M. A.; Wagoner, J. A. Fertilization Rate and Irrigation Scheduling in the Nursery Influence Growth, Insect Performance, and Stress Tolerance of 'Sutyzam' Crabapple in the Landscape. *HortScience* **2006,** *41* (2), 442—445.

51. Nielsen, D. G. Exploiting Natural Resistance as a Management Tactic for Landscape Plants. *The Florida Entomologist* **1989,** *72* (3), 413—418.

52. Debona, D.; Rodrigues, F. A.; Datnoff, L. E. Silicon's Role in Abiotic and Biotic Plant Stress. *Annual Review of Phytopathology* **2017,** *55,* 85—107.

53. Hong, S. C.; Williamson, R. C.; Held, D. W. Leaf Biomechanical Properties as Mechanisms of Resistance to Black Cutworm (Agrotis Ipsilon) Among *Poa* Species. *Entomologia Experimentalis et Applicata* **2012,** *145,* 201—208.

54. Raupp, M. J. Effects of Leaf Toughness on Mandibular Wear of the Leaf Beetle, *Plagiodera versicolora. Ecological Entomology* **1985,** *10,* 73-39.

55. McLarnon, E.; McQueen-Mason, S.; Lenk, I.; Hartley, S. E. Evidence for Active Uptake and Deposition of Si-Based Defenses in Tall Fescue. *Frontiers of Plant Science* **2017,** *8,* 1199.

56. Cacique, A. S.; Dominciano, G. P.; Moreira, W. R.; Rodrigues, F. A.; Cruz, M. F. A.; Serra, N. S.; Català, A. B. Effect of Root and Leaf Applilcations of Soluble Silicon on Blast Development in Rice. *Bragantia Campinas* **2013,** *72* (3), 304—309.

57. Korndorfer, A. P.; Cherry, R.; Nagata, R. Effect of Calcium Silicate on Feeding and Development of Tropical Sod Webworms (Lepidoptera: Pyralidae). *Florida Entomologist* **2004,** *87,* 393—395.

58. Redmond, C. T.; Potter, D. A. Silicon Fertilization Does Not Enhance Creeping Bentgrass Resistance to Black Cutworms or White Grubs. *Applied Turfgrass Science* **2006;** https://doi.org/10.1094/ATS-2006-1110-01-RS.

59. Goussain, M. M.; Moraes, J. C.; Carvalho, J. G.; Nogueira, N. L.; Rossi, M. L. Efeito da aplicação de silício em plantas de milho no desenvolvimento biológico da lagarta-Do-cartucho *Spodoptera frugiperda* (J.E.Smith) (Lepidoptera: Noctuidae). *Neotropical Entomology* **2002,** *31* (2), 305—310.

60. Hogendorp, B. K.; Cloyd, R. A.; Swiader, J. M. Effect of Silicon-Based Fertilizer Applications on the Reproduction and Development of the Citrus Mealybug (Hemiptera: Pseudococcidae) Feeding on Green Coleus. *Journal of Economic Entomology* **2009,** *102* (6), 2198—2208.

61. Assis, F. A.; Moraes, J. C.; Assis, G. A.; Parolin, F. J. T. Induction of Caterpillar Resistance in Sunflower Using Silicon and Acibenzolar-S-Methyl. *Journal of Agricultural Science and Technology A* **2015,** *17,* 543—550.

62. Ranger, C. M.; Singh, A. P.; Frantz, J. M.; Cañas, L.; Locke, J. C.; Reding, M. E.; Vorsa, N. Influence of Silicon on Resistance of *Zinnia elegans* to *Myzus persicae* (Hemiptera: Aphididae). *Environmental Entomology* **2009,** *38* (1), 129—136.

63. Brandhorst-Hubbard, J. L.; Flanders, K. L.; Appel, A. G. Oviposition Site and Food Preference of the Green June Beetle (Coleoptera: Scarabaeidae). *Journal of Economic Entomology* **2001,** *94,* 628—633.

64. Potter, D. A.; Held, D. W.; Rogers, M. E. Natural Organic Fertilizers as a Risk Factor for *Ataenius spretulus* Infestation on Golf Courses. *International Turfgrass Society Research Journal* **2005,** *10,* 753—760.

65. Lehmann, J.; Rillig, M. C.; Thies, J.; Masiello, C. A.; Hockaday, W. C.; Crowley, D. Biochar Effects on Soil Biota-A Review. *Soil Biology and Biochemistry* **2011,** *43,* 1812—1836.

66. Cook, S. P.; Neto, V. R. Laboratory Evaluation of the Direct Impact of Biochar on Adult Survival of Four Forest Insect Species. *Northwest Science* **2018,** *92* (1), 1—8.

67. Coy, R. M.; Held, D. W.; Kloepper, J. W. Rhizobacterial Inoculants Increase Root and Shoot Growth in 'Tifway' Hybrid Bermudagrass. *Journal of Environmental Horticulture* **2014,** *32,* 149—154.

68. Coy, R. M.; Held, D. W.; Kloepper, J. W. Bacterial Inoculant Treatment of Bermudagrass Alters Ovipositional Behavior, Larval and Pupal Weights of the Fall Armyworm (Lepidoptera: Noctuidae). *Environmental Entomology* **2017,** *46,* 831—838.

69. Coy, R. M.; Held, D. W.; Kloepper, J. W. Rhizobacterial Treatments of Tall Fescue and Bermudagrass Increases Tolerance to Damage from White Grubs. *Pest Management Science* **2019;** https://doi.org/10.1002/ps.5439. In press.

70. Coy, R. M.; Held, D. W.; Kloepper, J. W. Rhizobacterial Colonization of Bermudagrass by *Bacillus* Spp. In a Marvyn Loamy Sand Soil. *Applied Soil Ecology* **2019,** *141,* 10—17.

71. Coy, R. M.; Held, D. W.; Kloepper, J. W. Rhizobacterial Treatment of Bermudagrass Alters Tolerance to Damage from Tawny Mole Crickets (Neoscapteriscus vicinus Scudder). *Pest Management Science* **2019.** In press.

72. Frank, S. D.; Ranger, C. M. Developing a Media Moisture Threshold for Nurseries to Reduce Tree Stress and Ambrosia Beetle Attacks. *Environmental Entomology* **2016,** *45* (4), 1040—1048.

73. Potter, D. A.; Timmons, G. M. Flight Phenology of the Dogwood Borer (Lepidoptera: Sesiidae) and Implications for Control in *Cornus florida* L. *Journal of Economic Entomology* **1983,** *76,* 1069—1074.

74. White, T. C. R. The Abundance of Invertebrate Herbivores in Relation to the Availability of Nitrogen in Stressed Food Plants. *Oecologia* **1984,** *63,* 90—105.

75. Koricheva, J.; Larsson, S.; Haukioja, E. Insect Performance on Experimentally Stressed Woody Plants: A Meta-Analysis. *Annual Review of Entomology* **1998,** *43,* 195—216.

76. Pimentel, D. Insect Population Responses to Environmental Stress and Pollutants. *Environmental Reviews* **1994,** *2,* 1—5.

77. Huberty, A. F.; Denno, R. F. Plant Water Stress and its Consequences for Herbivorous Insects: a New Synthesis. *Ecology* **2004,** *85* (5), 1383—1398.

78. Coy, R. M. *Plant Growth-Promoting Rhizobacteria (PGPR) Mediate Interactions between Abiotic and Biotic Stresses in Cool- and Warm-Season Grasses.* PhD dissertation; Auburn University, 2017.

79. Zvereva, E. L.; Kozlov, M. V. Responses of Terrestrial Arthropods to Air Pollution: a Meta-Analysis. *Environmental Science and Pollution Research* **2010,** *17,* 297—311.

80. Larsson, S. Stressful Time for the Plant Stress-Insect Performance Hypothesis. *Oikos* **1989,** *56,* 277—283.

81. Staley, J. T.; Mortimer, S. R.; Masters, G. J.; Morecroft, M. D.; Brown, V. K.; Taylor, M. E. Drought Stress Differentially Affects Leaf-Mining Species. *Ecological Entomology* **2006,** *31,* 460—469.

82. Kettlewell, H. B. D. The Phenomenon of Industrial Melanism in Lepidoptera. *Annual Review of Entomology* **1961,** *6,* 245—262.

83. Cook, L. M.; Saccheri, I. J. The Peppered Moth and Industrial Melanism: Evolution of a Natural Selection Case Study. *Heredity* **2013**, *110*, 207–212.

84. Alstad, D. N.; Edmunds, G. F. Effects of Air Pollutants on Insect Populations. *Annual Review of Entomology* **1982**, *27*, 369–384.

85. Butler, C. D.; Beckage, N. E.; Trumble, J. T. Effects of Terrestrial Pollutants on Insect Parasitoids. *Environmental Toxicology & Chemistry* **2009**, *28* (6), 1111–1119.

86. Blande, J. D.; Holopainen, J. K.; Niinemets, U. Plant Volatiles in Polluted Atmospheres: Stress Responses and Signal Degradation. *Plant, Cell and Environment* **2014**, *37*, 1892–1904.

87. Anderson, W. F.; Snook, M. E.; Johnson, A. W. Flavonoids of Zoysiagrass (*Zoysia* spp.) Cultivars Varying in Fall Armyworm (*Spodoptera frugiperda*) Resistance. *Journal of Agricultural and Food Chemistry* **2007**, *55*, 1853–1861.

88. Berenbaum, M. Coumarins and Caterpillars: a Case for Coevolution. *Evolution* **1983**, *37*, 163–179.

89. Stoepler, T. M.; Lill, J. T. Direct and Indirect Effects of Light Environment Generate Ecological Trade-Offs in Herbivore Performance and Parasitism. *Ecology* **2013**, *94* (10), 2299–2310.

90. Rowe, W. J.; Potter, D. A. Shading Effects on Susceptibility of *Rosa* Spp. To Defoliations by *Popillia japonica* (Coleoptera: Scarabaeidae). *Environmental Entomology* **2000**, *29* (3), 502–508.

91. Kaur, N.; Gillett-Kaufman, J. L.; Gezan, S. A.; Buss, E. A. Association Between *Blissus insularis* Densitis and St. Augustinegrass Lawn Parameters in Florida. *Crop Forage Turfgrass Manage* **2016**, *2*. https://doi.org/10.2134/cftm2016.0015.

92. Folger, P. *Drought in the United States: Causes and Current Understanding.* Congressional Research Service Report 7-5700, 2017; p 25.

93. Ranger, C. M.; Reding, M. E.; Schultz, P. B.; Oliver, J. B. Influence of Flood-Stress on Ambrosia Beetle Host Selection and Implications for Their Management in a Changing Climate. *Agricultural and Forest Entomology* **2013**, *15*, 56–64.

94. Beck, E. W. Observations on the Biology and Cultural-Insecticidal Cointrol of *Prosapia bicincta,* a Spittlebug, on Coastal Bermudagrass. *Journal of Economic Entomology* **1963**, *56* (6), 747–752.

95. Martin, R. M.; Cox, J. R.; Alston, D. G.; Ibarra, F. Spittlebug (Homoptera: Cercopidae) Life Cycle on Bufflegrass in Northwestern Mexico. *Annals of the Entomological Society of America* **1995**, *88* (4), 471–478.

96. Peck, D. C. Seasonal Fluctuation and Phenology of *Prosapia* Spittlebugs (Homoptera: Cercopidae) in Upland Pastures of Costa Rica. *Environmental Entomology* **1999**, *28* (3), 372–386.

97. Ringenberg, R.; Lopes, J. R. S.; Müller, C.; Sampaio de Azvedo-Filho, W.; Paranhos, B. A. J.; Botton, M. Survey of Potential Sharpshooter and Spittlebug Vectors of *Xyllela fastidiosa* to Grapevines at the São Francisco River Valley, Brazil. *Revista Brasileira de Entomologia* **2014**, *58* (2), 212–218.

98. Blackshaw, R. P.; Coll, C. Economically Important Leatherjackets of Grassland and Cereals: Biology, Impact and Control. *Integrated Pest Management Reviews* **1999**, *4*, 143–160.

99. Potter, D. A. Influence of Feeding by Grubs of the Southern Masked Chafer on Quality and Yield of Kentucky Bluegrass. *Journal of Economic Entomology* **1982**, *75*, 21–24.

100. Gibb, T. J.; Bledsoe, L. W. Population and Behavior Modifications of Selected Arthropods Pests in Indiana during the 1988 Drought. *Proceedings of the Indiana Academy of Science* **1988**, *98*, 201–206.

101. Smitley, D. R.; Peterson, N. C. Twospotted Spider Mite (Acari: Tetranychidae) Population Dynamics and Growth of *Euonymus alata* 'Compacta' in Response to Irrigation Rate. *Journal of Economic Entomology* **1991**, *84* (6), 1806−1811.

102. Gillman, J. H.; Rieger, M. W.; Dirr, M. A. Drought Stress Increases Densities but Not Populations of Two-Spotted Spider Mite on Buddleia Davidii 'Pink Delight'. *HortScience* **1999**, *34* (2), 280−282.

103. Smitley, D. R.; Peterson, N. C. Interactions of Water Stress, Honeylocust Spider Mites (Acari: Tetranychidae), Early Leaf Abscission, and Growth of *Gleditsia tricanthos. Journal of Economic Entomology* **1996**, *89* (6), 1577−1581.

104. Kramer, K.; Cranshaw, W. S. Effects of Supplemental Irrigation on Populations of Clover Mite, *Bryobia praetiosa* Koch (Acari: Tetranychidae), and Other Arthropods in a Kentucky Bluegrass Lawn. *Southwestern Entomologist* **2009**, *34* (1), 69−74.

105. Ramirez, C. C.; Verdugo, J. A. Water Availability Affects Tolerance and Resistance to Aphids but Not the Trade-Off between the Two. *Ecological Research* **2009**, *24,* 881−888.

106. Cregg, B. M.; Dix, M. E. Tree Moisture and Insect Damage in Urban Areas in Relation to Heat Island Effects. *Journal of Arboriculture* **2001**, *27* (1), 8−17.

107. Hanks, L. M.; Denno, R. F. Natural Enemies and Plant Water Relations Influence the Distribution of an Armored Scale Insect. *Ecology* **1993**, *74* (4), 1081−1091.

108. Cockfield, S. D.; Potter, D. A. Interaction of Euonymus Scale (Homoptera: Diaspididae) Feeding Damage and Severe Water Stress on Leaf Abscission and Growth of *Euonymus fortunei. Oecologia* **1986**, *71,* 41−46.

109. Schuch, U. K.; Redak, R. A.; Bethke, J. A. Cultivar, Fertilizer, and Irrigation Affect Vegetative Growth and Susceptibility of chrysanthemum to Western Flowers Thrips. *Journal of the American Society for Horticultural Science* **1998**, *123* (4), 727−733.

110. Connor, E. F. Plant Water Deficits and Insect Responses: the Preference of *Corythuca arcuata* (Hetereoptera: Tingidae) for the Foliage of White Oak, *Quercus alba. Ecological Entomology* **1988**, *13,* 375−381.

111. Helmberger, M. S.; Craig, T. P.; Itami, J. Effects of Drought Stress on Oviposition Preference and Offspring Performance of the Lace Bug *Corythuca marmorata* on its Goldenrod Host, *Solidago altissima. Entomologia Experimentalis et Applicata* **2016**, *160,* 1−10.

112. Vazquez, J. C.; Buss, E. A. Southern Chinch Bug Feeding Impact on St. Augustinegrass Growth under Different Irrigation Regimes. *Applied Turfgrass Science* **2006**; https://doi.org/10.1094/ATS-2006-0711-01-RS.

113. White, T. C. R. Are Outbreaks of Cambium-Feeding Beetles Generated by Nutritionally Enhanced Phloem of Drought Stressed Trees? *Journal of Applied Entomology* **2015**, *139,* 567−578.

114. Jactel, H.; Petit, J.; Desprez-Loustau, M.-L.; Delzon, S.; Piou, D.; Battisti, A.; Koricheva, J. Drought Effects on Damage by Forest Insects and Pathogens: A Meta-Analysis. *Global Change Biology* **2012**, *18,* 267−276.

115. Kolb, T. E.; Fettig, C. J.; Ayers, M. P.; Bentz, B. J.; HIcke, J. A.; Mathiasen, R.; Stewart, J. E.; Weed, A. S. Observed and Anticipated Impacts of Drought on Forest Insects and Diseases in the United States. *Forest Ecology and Management* **2016**, *380,* 321−334.

116. Caldeira, M.; Fernandéz, V.; Tomé, J.; Pereira, J. S. Positive Effect of Drought on Longicorn Borer Larval Survival and Growth on eucalyptus Trunks. *Annals of Forest Science* **2002**, *59,* 99−106.

117. Hanks, L. M.; Paine, T. D.; Millar, J. G.; Campbell, C. D.; Schuch, U. K. Water Relations of Host Trees and Resistance to the Phloem-Boring Beetle *Phoracantha semipunctata* F. (Coleoptera: Cerambycidae). *Oecologia* **1999**, *119,* 400−407.

118. Hanks, L. M.; Paine, T. D.; Millar, J. G.; Hom, J. L. Variation Among Eucalyptus Species in Resistance to eucalyptus Borer in Southern California. *Entomologia Experimentalis et Applicata* **1995**, *74,* 185–194.

119. Hanks, L. M.; Paine, T. D.; Millar, J. G. Mechanisms of Resistance in Eucalyptus against Larvae of the eucalyptus Longicorn Borer (Coleoptera: Cerambycidae). *Environmental Entomology* **1991**, *20,* 1583–1588.

120. Heijari, J.; Nerg, A.-M.; Holopainen, J. K.; Kainulainen, P. Wood Borer Performance and Wood Characteristics of Drought-Stressed Scots Pine Seedlings. *Entomologia Experimentalis et Applicata* **2010**, *137,* 105–110.

121. Haack, R. A.; Hérard, F.; Sun, J.; Turgeon, J. J. Longhorned Beetle and Citrus Longhorned Beetle: a Worldwide Perspective. *Annual Review of Entomology* **2010**, *55,* 521–546.

122. Chakraborty, S.; Whitehill, J. G. A.; Hill, A. L.; Opiyo, S. O.; Cipollini, D.; Herms, D. A.; Bonello, P. Effects of Water Availability on Emerald Ash Borer Larval Performance and Phloem Phenolics of Manchurian and Black Ash. *Plant, Cell and Environment* **2014**, *37,* 1009–1021.

123. Muilenburg, V.; Herms, D. A. A Review of Bronze Birch Borer (Coleoptera: Buprestidae) History, Ecology, and Management. *Environmental Entomology* **2012**, *41* (6), 1372–1385.

124. Coleman, T. W.; Grulke, N. E.; Daly, M.; Godinez, C.; Schilling, S. L.; Riggan, P. J.; Seybold, S. J. Coast Live Oak, *Quercus agrifolia*, Susceptibility and Response to Gold-spotted Oak Borer, *Agrilus auroguttatus*, Injury in Southern California. *Forest Ecology and Management* **2011**, *261,* 1852–1865.

125. Rogers, L. E.; Grant, J. E. Infestation Levels of Dogwood Borer (Lepidoptera: Sesiidae) Larvae on Dogwood Trees in Selected Habitats in Tennessee. *Journal of Entomological Science* **1990**, *25* (3), 481–485.

126. Solomon, J. D.; Neel, W. W. Fecundity and Oviposition Behavior in the Carpenterworm, *Prionoxystus robiniae*. *Annals of the Entomological Society of America* **1974**, *67* (2), 238–240.

127. N'Guessan, K. F.; Kébé, I. B.; Adiko, A. Seasonal Variations of the Population of *Eulophonotus myrmeleon* Felder (Lepidoptera: Cossidae) in the Sud-Bandama Region of Côte d'Ivoire. *Journal of Applied Biosciences* **2010**, *35,* 2251–2259.

128. Hahn, P. G.; Maron, J. L. Plant Water Stress and Previous Herbivore Damage Affect Insect Performance. *Ecological Entomology* **2018**, *43,* 47–54.

129. Goergen, G.; Kumar, P. L.; Sankung, S. B.; Togola, A.; Tamò, M. First Report of Outbreaks of the Fall Armyworm *Spodoptera frugiperda* (J E Smith) (Lepidoptera, Noctuidae), a New Alien Invasive Pest in West and Central Africa. *PLoS One* **2016**, *11* (10), e0165632.

130. Mody, K.; Eichenberger, D.; Dorn, S. Stress Magnitude Matters: Different Intensities of Pulsed Water Stress Produce Non-monotonic Resistance Responses of Host Plants to Insect Herbivores. *Ecological Entomology* **2009**, *34,* 133–143.

131. Meyer, S. T.; Roces, F.; Wirth, R. Selecting the Drought Stressed: Effects of Plant Stress on Intraspecific and Within-Plant Herbivory Patterns of the Leaf-Cutting Ant *Atta colombica*. *Functional Ecology* **2006**, *20,* 973–981.

132. Hale, B. K.; Herms, D. A.; Hansen, R. C.; Clausen, T. P.; Arnold, D. Effects of Drought Stress and Nutrient Availability on Dry Matter Allocation, Phenolic Glycosides, and Rapid Induced Resistance of Poplar to Two Lymantriid Defoliators. *Journal of Chemical Ecology* **2005**, *31* (11), 2601–2620.

133. Dunn, J. P.; Kimmerer, T. W.; Potter, D. A. The Role of Host Tree Condition in Attacks of White Oaks by the Twolined Chestnut Borer, *Agrilus bilineatus* (Weber) (Coleoptera: Buprestidae). *Oecologia* **1986**, *70,* 596–600.

134. McCullough, D. G.; Poland, T. M.; Cappaert, D. Attraction of the Emerald Ash Borer to Ash Trees Stressed by Girdling, Herbicide Treatment, or Wounding. *Canadian Journal of Forest Research* **2009**, *39*, 1331−1345.

135. Potter, D. A.; Timmins, M.; Gordon, F. C. Flatheaded Apple Tree Borer (Coleoptera: Buprestidae) in Nursery-Grown Red Maples: Phenology of Emergence, Treatment Timing, and Response to Stressed Trees. *Journal of Environmental Horticulture* **1988**, *61* (1), 18−22.

136. Loughrin, J. H.; Potter, D. A.; Hamilton-Kemp, T. R.; Byers, M. Role of Feeding-Induced Plant Volatiles in Aggregative Behavior of the Japanese Beetle (Coleoptera: Scarabaeidae). *Environmental Entomology* **1996**, *25* (5), 1188−1191.

137. Loughrin, J. H.; Potter, D. A.; Hamilton-Kemp, T. R. Volatile Compounds Induced by Herbivory Act as Aggregation Kairomones for the Japanese Beetle (*Popilla japnica* Newman). *Journal of Chemical Ecology* **1995**, *21* (10), 1457−1467.

138. Kreuger, B.; Potter, D. A. Does Early-Season Defoliation of Crabapple (*Malus* sp.) by Eastern Tent Caterpillar (Lepidoptera: Lasiocampidae) Induce Resistance to Japanese Beetles (Coleoptera: Scarabaeidae)? *Journal of Entomological Science* **2003**, *38* (3), 457−467.

139. McBride, S.; Appel, D. *Hypoxylon Canker of Oaks;* Texas AgriLife Extension Publication EPLP-030, 2016; p 4.

140. Potter, D. A.; Hartman, J. R. Susceptibility of Honeylocust Cultivars to *Thyronectria austro-americana* and Response of *Agrilus* Borers and Bagworms to Infected and Non-infected Trees. *Journal of Environmental Horticulture* **1993**, *11* (4), 176−181.

141. Whyte, G.; Howard, K.; Hardy, G. E. S. J.; Burgess, T. I. The Tree Decline Recovery Seesaw; A Conceptual Model of the Decline and Recovery of Drought Stressed Plantation Trees. *Forest Ecology and Management* **2016**, *370*, 102−113.

Chapter 5

Sampling insects and decision-making

Sampling (a.k.a scouting) is the only way to know the abundance, location, and timing of activity of insects and mites in urban landscapes. Chapter 2, introduced the terms density and distribution as important descriptions for populations. All sampling and eventual pest management decisions are made at the population level. Sampling techniques for pests and beneficial insects in urban landscapes are not expensive or difficult, yet they are not commonly used. This may be due to several reasons; outbreaks are not as conspicuous as problems with annual or perennial weeds, insecticides on average are less expensive than fungicides or herbicides, and plant professionals typically do not know how to sample. In urban landscapes, insects and mites can be found above the ground, in the soil, or inside the plant (e.g., leafminers, borers). In the following sections, you will learn the common methods for sampling arthropods in the landscape (Box 5.1). The sampling methods needed for certain types of insects or mites can be specific to a species or group. The resources section at the end of this chapter provides a list of sources for sampling equipment and links to online resources for sampling.

BOX 5.1 Common terminology used in sampling

Sampling unit: what part of the insect habitat is being sample.

Sampling techniques: the methods being used. Examples include

- **Convenience samples**: nonreplicated, often single observations of a pest or damage.
- **Count samples:** numbers of insects in a particular part of their habitat.
- **Knockdown samples:** using chemicals, heat, or jarring methods to detect arthropods.
- **Traps:** tools used to passively intercept or actively draw arthropods to a collection point where they are stored until they are checked.
- **Nets:** tools to collect insects and mites in flight or from plants.

Urban Landscape Entomology. https://doi.org/10.1016/B978-0-12-813071-1.00005-1

Sampling insects and mites in turfgrass

Turfgrass is sampled using nets, traps, counts, excavation, or disclosing techniques. When sampling insects and mites in turfgrass you will quickly gain an appreciation for arthropod diversity. Although turfgrass is not as diverse as an old growth forest, most sampling techniques used to collect arthropods above ground will still produce a diverse sample. You will likely need a hand lens and the ability to recognize the life stage of interest.

A sweep net is the most common net type used for sampling caterpillars, chinch bugs, mites, adult beetles, flies, and some natural enemies in turfgrass. A sweep net is relatively inexpensive and fits easily behind the seat in the work vehicle. Smaller arthropods like foliage-feeding mites can be collected in sweep nets but most will require a hand lens to find them in a sample. Sweep nets are swung from side to side through the grass and then inverted to observe what is present (Fig. 5.1). Timing of sweeps can also determine the abundance of insects captured in sweep net samples. For example, sweep samples for fall armyworms are more productive at 2–3 p.m. compared to midmorning or early evening sweep samples. When fall armyworm populations are high, sweep samples at any time of day will collect them. However, when populations are low, a midafternoon timing is more likely to be successful (1). Leaf blowers, common equipment for landscape management companies can be modified to suck small insects from turfgrass. The modifications are relatively simple (2). Since the modifications remove the blower from regular service, the cost of a second unit may limit the use of this method. The collection bag provided with most blowers is too large to easily find insects. A smaller bag or pair of panty hose (2) able to tolerate the high velocity air flow and a firm way to attach it to the blower are needed.

FIGURE 5.1 Demonstration on using a sweep net to sample turfgrass.

Several traps can be used in turfgrass to monitor insects that are active above ground. Traps of any type are deployed and checked routinely (weekly or biweekly). Rainfall or frequent irrigation can be detrimental to trapping in turfgrass and should be considered as part of the plan to service these traps. Of these, pitfall traps are the only ones I have actually seen used. A pitfall trap is a collection container either placed on the turf surface or inserted into a hole in the ground. They will collect almost all types insects (pests and predators) and spiders that walk along the grass surface. In Idaho, a golf course superintendent uses a pitfall trap to learn when billbugs were active (Fig. 5.2). The bottom of the trap will contain either a food attractant or a preserving\killing solution. Soapy water or diluted anti-freeze are common preserving solutions. A pitfall trap baited with a piece of hot dog (Fig. 5.3) can be used to monitor imported fire ants. This tube is placed on the grass surface for about 15−30 min to determine if ants are present and foraging. Ant foraging is needed for bait treatments to be successful. Foraging by ants can also determined with the potato chip test (Fig. 5.3). Baited pitfall traps or mounds counts are useful methods to determine if ant control methods are working. A linear pitfall trap (Fig. 5.2) can determine the surface activity of mole crickets and billbugs. These are sections of house guttering or pipe installed into the ground (see Resources section for instructions). They are the most labor-intensive trap to install and best for sites with severe, annual problems with either billbugs or

FIGURE 5.2 A golf course superintendent is checking a linear pitfall trap for billbugs.

FIGURE 5.3   A piece of hot dog, or a regular (not fat free) potato chip on the ground can be used to determine if imported fire ants are actively foraging.

mole crickets. For mole crickets, larger life stages are more likely to be detected in these traps (*3*). Low densities are difficult to detect in linear pitfall traps (*3*) and capture of 10 or more per week indicates a high population (*4*). Be aware that pitfall traps can capture almost anything walking on the surface. Arthropod bycatch, occasional mice and snakes may also be captured. Large numbers of imported fire ants or millipedes, for example, can fill the trap in a short period making it more difficult to detect other insects.

Pheromones are available for the common caterpillar pests and some pest beetle species of turfgrass. Pheromones are then added to various types of traps to make them more target specific. Pheromone-baited traps are simple to deploy and specific to adults of one genus or species. However, they are rarely used because they attract the adult life stages of moths and beetles (mostly) which are not the life stages that do damage in turfgrass. There have been a limited number of attempts to relate moth counts from pheromone-baited traps to caterpillars in turfgrass with no significant correlations (*5*). Blacklight traps are nonspecific traps used to collect adult, flying insects like beetles especially when no specific attractant like a pheromone has been commercially produced. Blacklight traps can attract many different insects including many species that are not pests of turfgrass. They are also least effective during times when the moon is full.

Count sampling is the easiest sampling technique, but still rarely used. Counts are the only sampling technique to directly measure insects, and this provides the best estimate of pest population density. They are also the basis for treatment thresholds for turfgrass insects when known. In some instances, landscape and turfgrass managers use a count of one to trigger action against certain pests. Landscape managers in the southern United States treat whenever imported fire ant mounds are present either for esthetic reasons or to prevent clients from being stung. Excavation is needed to count insects in the soil like white grubs. A spade or cup cutter are used to take a plug of soil and turfgrass about 6—10 cm (2.5—4 inch) deep. The plug is then broken apart and the number of grubs counted.

Disclosing solutions use liquids to expose insects in soil, thatch, or within plants. The soap flush is used to sample turfgrass for caterpillars and mole crickets. A soap flush requires 30 mL (2 tablespoons) of liquid dishwashing soap mixed per 3.8 L (1 gallon) of water. This is best if mixed in a watering can so that the soap solution can be poured onto 0.1—0.2 m² (1—2 sqft) of infested turfgrass (Fig. 5.4). Within minutes, the insects, if present, will surface. There are anecdotes that lemon scented soap is the most effective for mole crickets, but this is not supported in the literature. Short and Koehler (6) evaluated many major brands of dishwashing soap and all were equally effective. A dilute solution of pyrethrins (1.2% pyrethrins plus piperonyl butoxide), 30 mL per 3.8 L of water, is actually more effective as a disclosing solution than all soap solutions (6). Both the pyrethroid and soap flush solutions will also cause earthworms and adult billbugs to emerge from soil if present, although there are no data confirming the optimal rate for those groups. Soil moisture at or above 20% is optimal and the ability to recover mole crickets and likely other pests decreases lower soil moisture (7). Practically speaking, 20% soil moisture is the point where the soil will remain a ball when squeezed. You should do this simple test before you invest time and resources attempting to do a soap flush for mole crickets. Because success depends on soil moisture, irrigation before sampling improves the chances of success. A second disclosing solution uses only water to sample chinch bugs and their predators. This flotation sample requires a can open on both ends to be pushed into the ground. The can is then filled with only water and the insects will float on the surface of the water.

FIGURE 5.4   A soap disclosing solution applied to turfgrass can sample mole crickets or turf-infesting caterpillars.

Another disclosing solution uses saltwater (1 cup of table salt per quart (0.95 L)) to sample larvae of billbugs and the annual bluegrass weevil inside the grass plant. Use a shovel or cup cutter to remove plugs of grass from an area suspected of being infested. Remove most of the soil from this plug and place it into a container deep enough so that the grass plug can be submerged in the salt solution. The plug will sit covered with saltwater for about 30−45 min. If larvae are present, they will float on the surface of the salt solution. The final disclosing solution is an alcohol wash used to sample grass-feeding mites (bermudagrass or zoysia grass mites). The type of alcohol (e.g., ethanol, isopropyl) is not important, and it can be diluted to 50% by volume with tap water. Similar to the saltwater technique, a plug of grass from a lawn suspected of being infested is taken. For this technique, only grass foliage is needed so it can be trimmed from the core using scissors. The grass foliage is then placed in a container with a liquid-tight lid and then alcohol is added. Shake the container with the grass and alcohol vigorously. The liquid should contain any mites that were present and will be green because of the extracted pigment from the grass. It would be easier to pour the liquid out into a white disposable bowl or plate with raised sides. Mites are difficult to see with the unaided eye so a hand lens is needed if mites are present.

Turfgrass can be sampled for damage using various rating scales. These approaches are indirect since they do not provide any estimate of the population but measure the aesthetic or horticultural impact of pests on plants. The simplest ratings are based on percentages (e.g., percent damage, percent of infested areas). Quality ratings that rank color and quality of a stand of grass are also used. A grid based system was developed at Auburn University to specifically measure mole cricket damage (8) (Fig. 5.5). The grid is a PVC

FIGURE 5.5   The Cobb and Mack (8) method of sampling damage by mole crickets. *Reprinted from Applied Soil Ecology, 94, D.L. Bailey, D.W. Held, A. Kalra, N. Twarakavi, F. Arriaga, Biopores from mole crickets (Scapteriscus spp.) increase soil hydraulic conductivity and infiltration rates, 7−14, 2015, with permission from Elsevier.*

pipe frame, 0.6 m (2 ft) square, that is divided into nine equal sections. You place the frame over an infested area and count the number of squares that have mole cricket damage. Each plot is ranked from 0 to 9 with 9 being the most damaged and 0 having no damage. This method is used to determine the effectiveness of insecticides to reduce damage caused by mole crickets. If an insecticide is effective, the damage rating will decrease over time after treatment.

Sampling insects and mites in ornamental plants

Similar types of sampling techniques used in turfgrass are also used for ornamentals. The same sweep net and modified leaf blower recommended for sampling turfgrass pests can be applied to sampling ornamental plants. However, a sweep net and even a vacuum sampler may be too forceful for new growth or some herbaceous ornamental plants. The same groups of insects, caterpillars, lace bugs, mites, leaf-feeding beetles, can be sampled with these techniques on ornamentals. Another method, called beat sampling, dislodges insects from plants by tapping branches over a sheet, sweep net, or other collection container (Fig. 5.6). Custom beat sheets, with wood supports, are used in beat sampling field crops. These are nice but require fast reflexes to collect the adult insects before they fly or are blown away. A nursery inspector in Alabama uses a plastic plate or disc toy for placing under a beat sample. A yellow- or white-colored plate or disc provides a better contrast, but insects can still either fly or blow away if they are not quickly collected. For these

FIGURE 5.6   Beat sampling azaleas to determine the presence and number of azalea lace bugs and predators.

reasons, a sweep net is still the better piece of equipment. You can sweep the foliage on stronger woody plants or tie the net portion in half to make it shallower for beat sampling tender vegetation. If you need to sample taller trees, the handle of the sweep net fits nicely into a 1.27–2.5 cm (0.5–1 inch) diameter PVC pipe to make an inexpensive extension pole. An alcohol wash discussed for sampling grass-feeding mites will also work for most smaller arthropods (thrips, mites, and aphids) on ornamental plants.

Ornamental insects are diverse, and so are traps used for these insects. Since traps only collect mobile life stages, most traps target the winged, adult life stage. Most traps must also have one or more lures to be effective. Lures can be visual or olfactory attractants that are specific (pheromones) (Box 5.2) or more generic (ethanol). This section will highlight the more common trap types, their target insects, and address the concerns about attracting pests to a location.

BOX 5.2 What are pheromones and kairomones?

Pheromones are volatile signals (smells) produced so individuals of the same species can communicate with one another. They have low molecular weights and are carried in the direction of prevailing winds. Insects of the same species "talk" using pheromones. Another signal confused with pheromones is kairomones. Kairomones are signals exploited by another organism for their benefit (9). Plants release kairomones that are exploited by certain insects.

Types of pheromones:
- *Sex or mating pheromones*: These are usually produced by females to attract males. They are relatively specific and used for monitoring adult insects.
- *Aggregation pheromones*: Some insects use these pheromones to form large clusters.
- *Trail pheromones*: This is how ants follow one another. One ant finds a food resource then applies a trail pheromone to the ground on the return trip to the colony. Other ants then follow that chemical trail to the resource. Trail pheromones can persist overnight enabling ants to find the resource again the next day.
- *Alarm pheromones*: These pheromones trigger defensive responses. For example, ants swarming from their mound or aphids dropping from a plant in response to a threat.

Examples of kairomones:
- Adult *Tiphia* wasps using the smell of grub frass (poop) to find white grubs underground (10).
- A plant producing a signal that attracts a natural enemy.
- A plant producing a signal when damaged by insects that attracts more of the same insect.
- Phorid flies use fire ant alarm pheromone and components of the venom to find ants in which to lay eggs (11).

Traps using pheromones. Traps that use sex pheromones target moths or beetles are usually specific. Almost any type of trap can be baited with pheromones. Japanese beetles and moths with wood-boring larval stages are the primary targets for traps in urban landscapes. There are justified and unjustified concerns about using highly effective traps for recruiting plant-feeding insects in urban environments. There is no concern about the recruitment of male moths to pheromone-baited traps. Adult moths are not the damaging life stage and males do not lay eggs that could locally increase the population of larvae. However, the recruitment and eventual spillover from traps is possible when using a generic attractant (like ethanol) or recruiting pests where the adult is the damaging. For example, some home improvement stores and garden centers sell "beetle bags" for Japanese beetles. However, there can be significant over recruitment of beetles to these traps. This causes incidental landings by beetles on plants on their way to the trap (*12*). And, since both male and female beetles feed, it can increase damage to adult host plants near the trap (*13*). I like to tell Master Gardeners that a Japanese beetle trap is the best Memorial Day gift you could ever give to your neighbor! This is one of the better documented cases of spillover with plant-feeding insects, but it is likely with other traps that attract adults and especially if both sexes do damage.

Other traps. Sticky traps, commonly yellow or white (Fig. 5.7), use color as the attractant. Yellow and white are the two most attractive colors for day-active insects. Greenhouse growers know the value of using these traps for small insects like whiteflies and thrips, but that technology has not translated

FIGURE 5.7 Yellow sticky cards can be deployed on stakes (A) in flower beds or hung from clips or wires in trees and shrubs (B).

to outdoor ornamentals. Sticky traps are some of the most easily deployed traps. They can be stuck to a wooden plant stake (Fig. 5.7A) or hung from a wire or loop from a branch in a tree (Fig. 5.7B). Sticky cards could detect the early arrivals of small pests like thrips, aphids or whiteflies to the landscape before populations cause damage. They are also useful to detect the presence of the adult life stage of some natural enemies.

Perhaps the most unusual trap for a landscape pest is the use of beer for slugs. Commercially available slug traps (e.g., Slug Saloon) are inserted into the ground and baited with a mixture of malted barley, rice, yeast, and sugar bait. Beer is also effective when used in these traps (14,15). The attraction to beer is by smell and not by taste (16) and attraction can be reduced if the beer is flat (14). Beers vary in their attractiveness to slugs and snails (15). Designs for homemade traps exist online but not all are effective. Hagnell et al. (15) found the online plastic bottle trap design was ineffective at capturing slugs relative to the commercial trap or a homemade box traps inserted in-ground trap (15).

Plant damage and Diagnostics

Researchers commonly use plant injury as an indirect indicator of insect and mite populations. Using plant symptoms or signs of pests requires knowing about the biology including how mouthparts relate to damage symptoms. Insects and their relatives in urban landscapes have different mouthparts (Table 5.1). To make this more confusing, adults and the larvae stages (but not nymphs) of the same species may have different mouthparts. This is true of moths and butterflies (Lepidoptera) and maggots. Insects with chewing mouthparts cause a loss of plant tissue as they feed, which is the most conspicuous type of plant damage. The appearance of the chewed leaves only provides limited insight about the insect involved. Leaf beetles and weevils generally make irregularly shaped holes in leaves that do not necessarily

TABLE 5.1 Types of mouthparts on plant-feeding insects, mites, and molluscs in urban landscapes.

Mouthparts	Common names
Chewing	White grubs, wood borers, leaf-feeding beetles, sawflies, caterpillars, ants, maggots, millipedes
Chewing (radula)	Slugs and snails (molluscs)
Sap-sucking	True bugs, aphids, whiteflies, scale insects, thrips, spider mites

originate from the leaf edges. Flea beetle adults are a common cause of shot hole damage in leaves. Shot hole damage looks like numerous small holes in the leaf as if shot by shotgun pellets. Some fungal pathogens can also cause this symptom. If insects are the cause of the shot hole damage, the insects or insect frass (solid insect waste) may also be present on the leaves. Scarab beetles are another leaf feeding beetle as adults and species are either nocturnal or diurnal. Diurnal scarab species, rose chafers (*Macrodactylus* spp.) and Japanese beetles (*Popillia japonica*), are colorful and will be obvious when present. Damage from the nocturnal scarabs is less common and damage symptoms appear quickly, perhaps in one evening. May/June beetles, for example, typically feed until just the petiole and midvein of the leaf remain, although this damage is not unique to these beetles. Damage happens quickly so once it is discovered, it often too late to respond. That type of damage happening overnight, especially when adjacent to an outdoor security light suggests May/June beetles.

Some chewed leaves can be caused by beneficial insects or slugs and snails. For example, leafcutter bees will cut almost perfect circular holes from the edges of leaves and flowers of many different plants (Fig. 5.8). Commonly, leaves on redbud trees, roses, Virginia creeper, or clematis show these circular

FIGURE 5.8   Leafcutter bees make almost perfect circular holes in the edges of certain land-scapes plants. The damage never causes defoliation or warrants management.

holes, but you will rarely see the bee. Weevils, snouted beetles, may also make circular holes along the leaf edge. However, holes made by weevils are not nearly as uniform in size and appearance as those made by leafcutter bees. Holes in leaves may also be caused by molluscs (snails and slugs). The mouthparts are a tongue with teeth called a radula. The radula teeth produce unusual feeding patterns best seen when they scrape algae from hard surfaces (Fig. 5.9A). There are several species of snails and slugs that feed on plants. Feeding results in holes in leaves that may resemble that of leaf-feeding insects. Often associated with these holes are mucus (slime trails), which dries into the silvery lines seen near the holes (Fig. 5.9B). Insects and millipedes with chewing mouthparts will not have associated slime trails. It is uncommon to see the snail or slug because they often feed at night or on overcast, wet or humid days. However, leafy perennials like hostas can be severely damaged by slugs.

Among leaf-feeding insects on plants, caterpillars and sawfly larvae are the most common defoliators in urban landscapes. Since there are so many families and species, it is important to introduce some terms that will be in later chapters. Larval insects with a caterpillar form are broadly called eruciform larvae. This body form (Fig. 5.10) can be either the larvae of moths or butterflies (caterpillars, Order: Lepidoptera) or sawflies (Order: Hymenoptera). These larvae have a distinct head, three pairs of thoracic legs behind the head, and paired, fleshy legs on the abdominal segments called prolegs. In most cases, eruciform larvae with more than five pairs of prolegs are sawfly larvae. Those with five or fewer are true caterpillars. Furthermore, the base of the prolegs where it touches the plant has sets of small hooks called crochets that look like eyelashes. If crochets are present, it is a caterpillar (Lepidoptera).

FIGURE 5.9 Slugs have a radula which makes unique feeding patterns (A). When feeding on plants, this produces holes associated with slime trails (B).

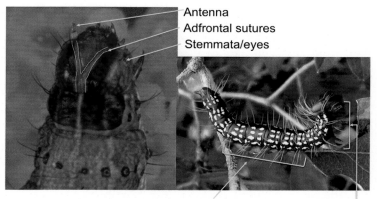

Antenna
Adfrontal sutures
Stemmata/eyes

Prolegs; Crochets, if present,
are located on the underside

True legs\
thoracic legs

FIGURE 5.10 Caterpillars and eruciform larvae have three pairs of true legs behind and prolegs on certain abdominal segments. The presence and appearance of crochets and the number of simple eyes are important secondary characteristics on these larvae.

Prolegs and crochets help eruciform larvae hold on to their host plant. If you have a good hand lens, the number of simple eyes (called stemmata) can be used. If more than one pair (one on each side of the head) of these simple eyes are present, then it is a caterpillar. Separation of eruciform larvae to order can be important because caterpillars and sawflies may share the same host plants. Also, chemical controls with novel modes of action (growth regulators or biologicals) may not have activity against all eruciform larvae. The most common example is the formulation of the *kurstaki* strain of the bacteria *Bacillus thuringiensis* (*Bt*). The crystal (*cry*) proteins that are effective against lepidopteran larvae are not the same ones effective against sawfly larvae. Some references incorrectly state that *Bt* does not work against sawflies but that is not entirely true. Some *Bt* strains can cause some mortality against sawflies and other insects in the order Hymenoptera (e.g., ants) (*17,18*); however, the strain of *Bt* in common insecticide formulations is not effective against sawflies (*19*).

Leafminer larvae all have chewing mouthparts. These insects will tunnel between the top and bottom layers of leaves (Fig. 5.11), leaving characteristic evidence of their feeding. The mines, tunnels inside the leaf left after feeding, are further characterized as linear, serpentine, or blotch mines. Linear and serpentine mines appear as straight or squiggled lines on the leaf. Blotch mines may or may not take on the empty appearance, but will be discolored. Numbers of mined leaves is an indirect measure of plant injury. Mines are not reused since the plant cells were already consumed, but leafmines remain as long as the leaf is alive. Similar to leafmines, wood-boring insects create

FIGURE 5.11 Leafminers are larvae that tunnel between the leaf layers. The damage shown is from boxwood leafminer, a type of blotch miner.

galleries inside the main trunk or stems (Fig. 5.12). Like leafminers, galleries can have unique patterns and be used for pest identification.

Sap-sucking is a generic term that describes insects and mites that consume xylem or phloem fluid, or plant cell contents. Insects that use xylem, for example, need powerful sucking muscles to work against the negative fluid pressure in that vascular tissue. It is usually apparent that an insect is a xylem feeder if it has an unusually large head (Fig. 5.13). Essentially without that muscle, the negative pressure in the xylem would pull fluids from the insect rather than vice versa. The signs and symptoms of insects that feed on plant fluids are subtle since they do not consume the leaf blade. The most common symptoms are "rain" or mistlike droplets, lerps or spittle, stippling, and honeydew or sooty mold. Rain or mist, honeydew, spittle, and lerps are liquid excrement projected (thus the name sharpshooters) from insects that consume large volume of either xylem or phloem. Rain, mist, and honeydew drops from plants onto people, plants, and object below. Persons under trees infested with leafhoppers, sharpshooters, or certain aphids or scale insects report feeling an occasional mist or drop, and may use the phrase, "crying tree" as a description. Spittle masses on woody or herbaceous ornamentals or turfgrass are produced by nymphs of the superfamily Cercopoidea (froghoppers and spittlebugs). The spittle mass protects them from desiccation and from insect predators (20). Stippling results when an insect (e.g., leafhoppers or lace bugs) or mite consumes plant cell contents (Fig. 5.14). What remains are cells that lack chlorophyll (green pigment) and appear to have yellow spots, which are sometimes

FIGURE 5.12   Wood-boring insects create galleries inside host plants where the larvae develop.

confused with a nutrient deficiency. Under heavy insect pressure, the individual feeding spots can cover most of the leaf surface.

Honeydew is the term reserved for insects that use phloem (21). Honeydew is mostly sugar with amino acids, and therefore is sticky. Insects that produce honeydew have adaptations like waxy coatings to prevent them from becoming stuck to surfaces where honeydew lands. The lerp of lerp psyllids is a mixture

FIGURE 5.13   The glassy-winged sharpshooter has a bulging head common of insects that feed on xylem.

FIGURE 5.14 Stippling is caused by the removal of cell contents by certain sap-sucking insects. This is sometimes confused with a nutrient deficiency.

of waxes and honeydew that form a crystallized domicile in which the immature psyllids are protected during development (*21*). Once on surfaces, honeydew may partially dry to make the surface shiny, but it remains sticky. On sidewalks and grass under trees infested with certain sap-sucking insects, the shiny and sticky surfaces may stick to the soles of shoes. The shiny drops often go unnoticed or are easily dismissed as rain on leaves or a car windshield. However, it is difficult to ignore once soot mold fungi spores land and begin to grow (*22*). Sooty mold fungi have dark mycelium (Fig. 5.15), like a darker version of bread mold except it tightly adheres to branches and leaves and persists from year to year. Once established, sooty mold can be difficult to remove even with soap and water and even pressure washing (*22,23*). One application of a dilute (1%) solution of oil soap, but not insecticidal soap, reduces sooty mold cover on leaves of crape myrtle and was not phytotoxic when applied and left to dry on leaves of eight different woody plant species (*23*). In India, a darkling beetle was discovered that consumes and removes sooty mold from leaves (*24*). There may eventually be a biological solution for the problem of sooty mold. Until then, elimination of the sap-sucking insects will prevent additional honeydew and sooty mold from accumulating.

Webbing is commonly associated with spiders, but webbing or silk on woody or herbaceous ornamental plants can be produced by certain plant-feeding arthropods with sucking and chewing mouthparts. Heavy populations of spider mites can cover leaves or flowers with webbing (Fig. 5.16). Among silk-producing insects, barklice, webspinners, and caterpillars are

FIGURE 5.15 Sooty mold fungi form dark mats of growth on top of honeydew from phloem-feeding insects. Often sooty mold is mistaken for a plant infection, it only grows where honeydew is present (*22*).

commonly encountered in urban landscapes. Barklice are small aphidlike creatures with chewing mouthparts. They create thick silken mats on the trunk and branches of trees on which they move around the plant (Fig. 5.17). These mats can sometimes go from the ground into the upper parts of the trees. Webspinners create similar mats of silk but usually those silken mats are near the tree bole and in the adjacent mulch or leaf litter. Some caterpillars use silk to create individual bags or nests. Bagworms (Psychidae) form individual bags around a single caterpillar. Webworms, tent caterpillars, and nest-building sawflies have communal nests, a silken bag used by a group of larvae. Of these, eastern tent caterpillars form a nest in the central part of the tree compared to nests of webworms and sawflies that usually cover leaves at the end of a branch (Fig. 5.16).

Evaluations of plant varieties for susceptibility to pests usually summarize the impacts as either a damage rating, percent leaf damage, or percent defoliation. For borers, frass exuding from galleries or trunk wounds are counted (*25*). For some insects, the amount of plant injury is related to plant mortality, reduced plant esthetics, or market value. For example, observations suggest that 10 or more attacks (indicated by sawdust toothpicks) by the granulate ambrosia beetle on trees of 7.6 cm (3 inch) or less caliper will be fatal (*26*).

FIGURE 5.16 Plant-feeding insects and certain mites can produce silk. Eastern tent caterpillars (upper) produce nests on the interior of the tree, webworms (middle) make nest over leaves or branches where they feed. Spider mites (lower) produce silk usually when populations are high.

FIGURE 5.17   Small insects called barklice can produce large mats of webbing that can cover the main trunk and branches of the tree. This is common on live oak trees along the Gulf Coast in the United States. The inset image shows barklice using the silk to move around the plant. Unlike caterpillars, barklice do not feed on the plants that they cover with silk.

Consumers are willing to accept 7% or less defoliation of canna lilies by Japanese beetles in their landscape or a public garden (27). Surveyed home-owners indicated that >3.3% actual canopy injury to an azalea plant by azalea lace bug would be sufficient to prompt treatment for 50% of the survey participants (28). We will discuss the relationship of injury to treatment and perception of damage later in this chapter.

Phenology and degree days

In Chapter 4, you were introduced to the effects of temperature on insect development. We will now apply those ideas through the concepts of degree days and phenology. Temperature is measured in degrees but it can also be measured biologically through phenology, or the sequence of seasonal biological events. Each year as spring begins, daffodils and crocus emerge and bloom, and the woody shrub forsythia blooms. These two events are among the earliest in the phenological sequence throughout much of the continental United States. As the season progresses, each species plays its role through either leaf flush, bloom, or both. This sequence happens in the same manner from year to year. In urban landscapes, we fortunately have a wide range of flowering plants to provide a biological calendar (29) with phenology plotted over time. For example, turfgrass managers in the northeastern United States use the green and gold stage of bloom of *Forsythia intermedia* as a phenological indicator of activity of the annual bluegrass weevil (*30*). Plant phenology allows us to make predictions of when these events will occur based on events that would be obvious to people in plant sciences. However, these events are still based on temperature data and can also be calculated using degree days.

Degree days are an accumulation of heat during a 24-h period. This concept is widely taught in introductory Entomology courses and is analogous to chilling hour calculations taught in some Horticulture courses. Degree days are calculated values based on the measured daily high and low temperatures and the amount of heat that accumulates above some base temperature. There are different methods for calculating degree days (*29*) but all require daily temperature data, a date when to begin calculations (biofix), and a base temperature (usually 10°C [50°F]). In research publications, degree-day values are presented as a range of values representing accumulations from the biofix date. Degree-day calculations are not absolute, and most of the variation is due to the accuracy of the weather data being used. For example, if the weather station being used is at an airport, degree days based on those weather data will accumulate faster than weather data taken from a suburban or even rural weather station. The same effect of microclimates could happen with phenological indicators. Plants that are in the city center will be more phenologically advanced. For example, rural and urban areas around Madison, Wisconsin, can vary by 200 or more degree days (base 10°C [50°F]) in the same year. This also means the first freezing temperatures occur a few weeks earlier in rural areas, and spring can be 1−3 weeks earlier in urban areas (*31*). Because of microclimates, plant correlates and weather data for a pest insect or mite are more accurate if located closer to the property being managed.

For some of the important pests of turfgrass and ornamentals, there are published degree-day accumulations or plant phenological calendars. Herms (*29*) is a book chapter dedicated to degree models and phenology with examples of sequences from OH and MI. Dr. Herms was the first to do statewide phenological studies with ornamental plants in the early 1990s. Since then, similar studies have been repeated in Kentucky (*32*), Tennessee (*33*), and Alabama (*34*). If your specific state is not represented by these studies, data from the same USDA hardiness zone would be the next best option (*34*).

Assessing damage and making decisions

In field crops, pest management decisions are based on the economic injury levels and economic thresholds. Thresholds are the numbers of insects per plant or plant part where action is needed to prevent populations from reaching a level where damage or crop loss occurs. In urban landscapes there are no yield losses to measure and most common landscape pests will not kill their host plants. At best, we can only determine loss of esthetics or replacement value should a lawn or tree be killed. Loss of esthetics can be challenging. A retired colleague at Auburn University introduced me to the phrase "esthetically dead plant." This is a landscape plant that is weakened by biotic or abiotic stresses to a point where it no longer has esthetic or functional utility. This could be smaller flowers produced by crape myrtle trees infested with crape myrtle bark scale (*35*) or the defoliation of hedges used as living fences between properties. The models used to determine economic thresholds for crops are not capable of using bloom size or functional value unless the damage is tied to some dollar value. This was one of the driving reasons behind the development of the principles of an esthetic injury level a term coined by Olkowski (*36*). The idea was to develop an esthetic injury level that would function like an economic injury level. The esthetic injury level would have similar calculations but would require human respondents to determine the association between levels of injury and esthetic loss. Estimating damage based on esthetics is not easy. Values have been calculated successfully but not always successfully applied. Following the development of this idea, a series of papers (*28,37–40*) did the survey work and calculations for urban landscape pests. Common themes that emerged from this research on esthetic injury level calculations were as follows: (1) calculated values for esthetic injury levels are very low; (2) consumers were able to discern very small amounts of esthetic injury; and (3) costs of sprays are low, one-third, or less than the replacement costs of the infested plants. For example, the esthetic injury level for evergreen bagworm is about four larvae per 1.2 m (4 ft) tree (*37*). Since the early 1990s, few additional studies have calculated esthetic injury levels for pests, and the assumption that esthetic injury level is very low for ornamental plants and turfgrass has become generally accepted. That said, the context of the pest infestation in ornamentals can change the survey responses. It is clear from research and observations that people are more tolerant of pest damage in the landscape than at the point of purchase (*27*).

Despite the limitations of our current application of esthetic injury level in urban landscapes, the impacts of the original works of Olkowski (*37,41*) on urban landscape integrated pest management (IPM) have largely been underappreciated. Olkowski (*41*), for example, originally raised most of the same concerns addressed in this book. It seems the field is just now maturing into the vision outlined by Olkowski (*41*). The studies since the original esthetic injury level proposal recognize the high pesticide inputs in urban landscapes from the late 1960s–1980s and worked to apply ecological principles to reduce pesticide inputs. Table 5.2 summarizes selected published studies where insecticide (or pesticide) use was compared before and after implementation of an IPM program. These IPM programs still used insecticides but incorporated monitoring, biological controls, and cultural practices. Therefore, IPM is not intended to be synonymous with organic. IPM works to reduce pest populations through economically and ecologically sound practices that minimize hazards to humans, beneficial species, and the environment. The selected studies have two common outcomes: (1) labor costs usually increase with IPM due to scouting and (2) pesticide use is reduced over time when IPM is applied.

TABLE 5.2 Select studies demonstrating pesticide reductions through Integrated Pest Management (IPM) in landscapes.

Context or pest targeted	IPM components	Results
Orangestriped oakworm in Norfolk, Virginia (42)	• Established an esthetic injury level of 25% tree defoliation reduced the need for the city to spray on demand • Monitoring • Physical removal of egg masses • Using insecticides with reduced impacts on natural enemies	• Peak annual pesticide use against this pest was 55,172 L (14,345 gallons) • Year 1 of IPM program produced an 80% reduction in insecticide use for the pest • No insecticide use for this pest after 1999
Pilot program on five sites near Atlanta, GA (43)	• Biweekly scouting and IPM program including pests and beneficials • Growing degree days to predict pest activities • Beat samples	• Reductions in pesticide use 75%–99.3% over 2 years • Increase in lower toxicity insecticides • Labor costs increased but program costs decreased over 2 years below pre-IPM costs
Penn-Del IPM subscription service (44)	• Program • Growing degree days to predict pest activities • Plant phenological correlates	• Subscribers report 28%–41% reduction in pesticide use
Landscape IPM program in Montgomery Co., Maryland (45)	• Biweekly scouting of pests and beneficials • Implemented cultural controls in addition to insecticides	• Properties were on cover spray programs • 47%–63% reduction in pesticide use over the 2-year IPM program • Program costs reduced 12%–31% • Labor costs increased • Most (81%) of residents indicated plant aesthetics improved on IPM

TABLE 5.2 Select studies demonstrating pesticide reductions through Integrated Pest Management (IPM) in landscapes.—cont'd

Context or pest targeted	IPM components	Results
Homeowner and institutional IPM program in central Maryland (46)	• Biweekly scouting	• Decrease from 9 different insecticides and fungicides to 7 in 1 year • 75% reduction in variable costs on IPM • Increase in lower toxicity pesticides on IPM • 10 plant species accounted for 21% of all pest problems
Homeowner and institutional IPM program in central Maryland (47)	• 26 properties	• 94% reduction in pesticide use
Street tree IPM program in Berkley, CA (41)	• Applied esthetic injury level • Introduced natural enemies • Decisions made with an understanding of pest ecology	• 9 different insecticides used annually including 65 gallons of DDT • 3 insecticides in use after 3 years, 1 insecticide used after 5 years • Lower toxicity insecticides replaced more toxic materials

Highlights

- Sampling, although rarely practiced, is key to making IPM work effectively in urban landscapes.
- Degree-day accumulations using soil and air temperatures and plant phenology can predict the activity of certain pest life stages.
- IPM in urban landscape has two common outcomes: (1) labor cost increases with IPM due to scouting; (2) pesticide use is reduced by >40% when IPM is applied over time.

Resources

Selected sources for sampling equipment, traps, and supplies:

- Bioquip Inc, https://bioquip.com
- Gemplers Inc., https://www.gemplers.com
- Forestry suppliers https://www.forestry-suppliers.com
- Trécé http://trece.com

• Great Lakes IPM https://www.greatlakesipm.com

Selected sources for sampling and degree day information:

• TurfFiles: https://www.turffiles.ncsu.edu/insects/monitoring-for-turf-insects/
  - Database of information on deploying traps and sampling specifically in turfgrass.
• Syngenta Greencast Online: http://www.greencastonline.com/default.aspx.
  - Online soil temperature resource, degree-day calculator, and models for certain pests.

References

1. Alvarado, E. A.; Fuxa, J. R.; Wilson, B. H. Correlation of Absolute Population Estimates of *Spodoptera frugiperda* (Lepidoptera: Noctuidae) with Sweep Sampling and Yield in Bermudagrass. *Journal of Economic Entomology* **1983**, *76*, 792−796.
2. Zou, Y.; van Telgen, M. D.; Chen, J.; Xiao, H.; de Kraker, J.; Bianchi, F. J.; van der Werf, W. Modification and Application of a Leaf Blowervac for Field Sampling of Arthropods. *Journal of Visualized Experiments* **2016**, *114*, e54655. https://doi.org/10.3791/54655.
3. Hudson, W. G. Surface Movement of the Tawny Mole Cricket, *Scapteriscus vicinus* (Orthoptera: Gryllotalpidae). *International Turfgrass Society Research Journal* **2001**, *9*, 774−779.
4. Adjei, M. B.; Frank, J. H.; Gardner, C. S. Survey of Pest Mole Crickets (Orthoptera: Gryllotalpidae) Activity on Pasture in South-Central Florida. *Florida Entomologist* **2003**, *86* (2), 199−205.
5. Hong, S. C.; Williamson, R. C. Comparison of Sticky Wing and Cone Pheromone Traps for Monitoring Seasonal Abundance of Black Cutworm Adults and Larvae on Golf Courses. *Journal of Economic Entomology* **2004**, *97* (5), 1666−1670.
6. Short, D. E.; Koehler, P. G. A Sampling Technique for Mole Crickets and Other Pests in Turfgrass and Pasture. *Florida Entomologist* **1979**, *62* (3), 282−283.
7. Hudson, W. G. Field Sampling and Population Estimation of the Tawny Mole Cricket (Orthoptera: Gryllotalpidae). *Florida Entomologist* **1989**, *72*, 337−343.
8. Cobb, P. P.; Mack, T. P. A Rating System for Evaluating Tawny Mole Cricket, *Scapteriscus vicinus* Scudder, Damage (Orthoptera: Gryllotalpidae). *Journal of Entomology Science* **1989**, *24*, 142−144.
9. Murali-Baskaran, R. K.; Sharma, K. C.; Kaushal, P.; Kumar, J.; Parthiban, P.; Senthil-Nathan, S.; Mankin, R. W. Role of Kairomone in Biological Control of Crop Pests-A Review **2018**, *101*, 3−15.
10. Rogers, M. E.; Potter, D. A. Kairomones from Scarabaeid Grubs and Their Frass as Cues in Below-Ground Host Location by the Parasitoids *Tiphia vernalis* and *Tiphia pygidialis* **2002**, *102*, 307−314.
11. Sharma, K.; Vander Meer, R. K.; Fadamiro, H. Y. Phorid Fly, *Pseudacteon tricuspis*, Response to Alkylpyrazine Analogs of a Fire Ant, *Solenopsis invicta*, Alarm Pheromone. *Journal of Insect Physiology* **2011**, *57*, 939−944.
12. Switzer, P. V.; Enstrom, P. C.; Schoenick, C. A. Behavioral Explanations Underlying the Lack of Trap Effectiveness for Small-Scale Management of Japanese Beetles (Coleoptera: Scarabaeidae). *Journal of Economic Entomology* **2009**, *102* (3), 934−940.

13. Gordon, F. C.; Potter, D. A. Japanese Beetle (Coleoptera: Scarabaeidae) Traps: Evaluation of Single and Multiple Arrangements for Reducing Defoliation in Urban Landscape. *Journal of Economic Entomology* **1986,** *79,* 1381−1384.

14. Cranshaw, W. *Attractiveness of Beer and Fermentation Products to the Gray Garden Slug,* Agriolimax reticulatum *(Muller) (Mollusca: Limacidae);* Colorado State University Technical Bulletin TB97-1, 1997; p 7.

15. Hagnell, J.; Schander, C.; Nilsson, M.; Ragnarsson, J.; Valstar, H.; Wollkopf, A. M.; von Proschwitz, T. How to Trap a Slug: Commercial versus Homemade Slug Traps. *Crop Protection* **2006,** *25,* 212−215.

16. Piechowicz, B.; Grodzicki, P.; Piechowicz, I.; Stawarczyk, K. Beer as Olfactory Attractant in the Fight against Harmful Slugs Arion lusitanicus Mabille 1868. *Chemistry-Didactics-Ecology-Metrology* **2014,** *19,* 119−125.

17. Smirnoff, W. A.; Berlinguet, L. A Substance in Some Commercial Preparation of *Bacillus thuringiensis* Var. *thuringiensis* Toxic to Sawfly Larvae. *Journal of Invertebrate Pathology* **1966,** *8,* 376−381.

18. Garcia-Robles, I.; Sánchez, J.; Gruppe, A.; Martínez-Ramírez, A. C.; Rausell; Real, M. D.; Bravo, A. Mode of Action of *Bacillus thuringiensis* PS86Q3 Strain in Hymenopteran Forest Pests. *Insect Biochemistry and Molecular Biology* **2001,** *31,* 849−856.

19. Ibrahim, M. A.; Griko, N.; Junker, M.; Bulla, L. A. *Bacillus thuringiensis*: A Genome and Proteomic Perspective. *Bioengineered Bugs* **2010,** *1* (1), 31−50.

20. Nachappa, P.; Guillebeau, L. P.; Braman, S. K.; All, J. N. Susceptibility of Twolined Spittlebug (Hemiptera: Cercopidae) Life Stages to Entomophagous Arthropods in Turfgrass. *Journal of Economic Entomology* **2006,** *99* (5), 1711−1716.

21. Mittler, T. E.; Douglas, A. E. Honeydew. In *Encyclopedia of Insects;* Resh, V. H., Cardé, R. T., Eds.; Academic Press: Boston, Massachusetts, United States, 2003; pp 523−526.

22. Laemmlen, F. F. *Sooty Mold.* University of California Pest Notes #74108; University of California Agriculture and Natural Resources, 2011; p 3.

23. Held, D. W.; Wheeler, C.; McLaurin, W. Cultural Practices for Removal of Wax Scales and Sooty Mold from Ornamentals. In *Proceedings of the Southern Nursery Association Research Conference* **2006,** *51*; pp 141−144.

24. Josephrajkumar, A.; Mohan, C.; Poorani, J.; Babu, M.; Krishnakumar; Hedge, V.; Chowdappa, P. Discovery of a Sooty Mould Scavenging Beetle, *Leiochrinus nilgirianus* Kaszab (Coleoptera: Tenebrionidae) on Coconut Palms Infested by the Invasive Rugose Spiralling Whitefly, *Aleurodicus rugioperculatus* Martin (Hemiptera: Aleyrodidae). *Phytoparasitica* **2018,** *46,* 57−61.

25. Rogers, L. E.; Grant, J. E. Infestation Levels of Dogwood Borer (Lepidoptera: Sesiidae) Larvae on Dogwood Trees in Selected Habitats in Tennessee. *Journal of Entomological Science* **1990,** *25* (3), 481−485.

26. Mizell, R.; Riddle, T. C. Evaluation of Insecticides to Control the Asian Ambrosia Beetle, *Xylosandrus crassiusculus.* In *Proceedings of the Southern Nursery Association Research Conference* **2004,** *49*; pp 152−155.

27. Sadof, C. S.; Sclar, D. C. Public Tolerance to Defoliation and Flower Distortion in a Public Horticulture Garden. *Journal of Economic Entomology* **2002,** *95* (2), 348−353.

28. Klingeman, W. E.; Braman, S. K.; Buntin, G. D. Evaluating Grower, Landscape Manager, and Consumer Perceptions of Azalea Lace Bugs (Heteroptera: Tingidae) Feeding Injury. *Journal of Economic Entomology* **2000,** *93,* 141−148.

29. Herms, D. A. Using Degree-Days and Plant Phenology to Predict Pest Activity. In *IPM (Integrated Pest Management) of Midwest Landscapes;* Krischik, V., Davidson, J., Eds.; Minnesota Agricultural Experiment Station Publication 58-07645, 2004; pp 49–59.

30. Vittum, P. J.; Villani, M. G.; Tashiro, H. *Turfgrass Insects of the United States and Canad,* 2nd ed.; Cornell University Press: Ithaca, 1999.

31. Schatz, J.; Kucharik, C. J. Urban Heat Island Effects on Growing Seasons and Heating and Cooling Degree Days in Madison, Wisconsin USA. *International Journal of Climatology* **2016,** *36,* 4873–4884.

32. Mussey, G. J.; Potter, D. A. Phenological Correlations between Flowering Plants and Activity of Urban Landscape Pests in Kentucky. *Journal of Economic Entomology* **1997,** *90,* 1615–1627.

33. Klingeman, W. E.; Hoyt, K.; Flanagan, P.; Hale, F. Seasonal Pest Activity Patterns, Observations and Recommendations from a Tennessee Phenology Garden Pilot Study. *Proceedings of the Southern Nursery Association Research Conference* **2013,** *58,* 93–98.

34. Young, R. A. *Alabama Phenology Garden Project: Using Degree Days and Plant Phenology to Predict Pest Activity.* MS thesis; Auburn University, 2012.

35. Layton, B. *Crape Myrtle Bark Scale Identification and Control;* Mississippi State University Extension Publication 2838, 2016; p 8.

36. Olkowski, W. A Model Ecosystem Management Program. In *Proceedings of the Tall Timbers Conference Ecology and Animal Control and Habitat Management* **1974,** *5;* pp 103–117.

37. Raupp, M. J.; Davidson, J. A.; Koehler, C. S.; Sadof, C. S.; Reichelderfer, K. Decision-making Considerations for Aesthetic Damage Caused by Pests. *Bulletin of the Entomological Society of America* **1988,** *34,* 27–32.

38. Raupp, M. J.; Davidson, J. A.; Koehler, C. S.; Sadof, C. S.; Reichelderfer, K. Economics and Aesthetic Injury Levels and Thresholds for Insect Pests of Ornamental Plants. *Florida Entomologist* **1989,** *72,* 403–407.

39. Coffelt, M. A.; Schultz, P. B. Development of an Aesthetic Injury Level to Decrease Pesticide Use against Orangestriped Oakworm (Lepidoptera: Saturniidae) in an Urban Pest Management Project. *Journal of Economic Entomology* **1990,** *83,* 2044–2049.

40. Sadof, C. S.; Alexander, C. M. Limitations of Cost-Benefit-Based Aesthetic Injury Levels for Managing Twospotted Spider Mites (Acari: Tetranychidae). *Journal of Economic Entomology* **1993,** *86,* 1516–1521.

41. Olkowski, W.; Olkowski, H.; Van den Bosch, R.; Hom, R. Ecosystem Management: a Framework for Urban Pest Control. *BioScience* **1976,** *26,* 384–389.

42. Schultz, P. B.; Sivyer, D. B. An Integrated Pest Management Success Story: Orangestriped Oakworm Control in Norfolk, Virginia, U.S **2006,** *32,* 286–288.

43. Stewart, C. D.; Braman, S. K.; Sparks, B. L.; Williams-Woodward, J. L.; Wade, G. L.; Latimer, J. G. Comparing an IPM Pilot Program to a Traditional Cover Spray Program in Commercial Landscapes. *Journal of Economic Entomology* **2002,** *95* (4), 789–796.

44. Hoover, G. Collaborative for Integrated Pest Management. *Tree Care Industry* **2002,** *13,* 19–24.

45. Smith, D. G.; Raupp, M. J. Economic and Environmental Assessment of an Integrated Pest Management Program for Community-Owned Landscape Plants. *Journal of Economic Entomology* **1986,** *79,* 162–165.

46. Raupp, M. J.; Noland, R. M. Implementing Landscape Plant Management Programs in Institutional and Residential Settings. *Journal of Arboriculture* **1984,** *10* (6), 161–169.

47. Holmes, J. J.; Davidson, J. A. Integrated Pest Management for Arborists: Implementation of a Pilot Program in Maryland. *Journal of Arboriculture* **1984,** *10* (3), 65–70.

Chapter 6

Insects and mites in turfgrass

Turfgrass arthropod diversity is made of pests that utilize different habitats within the stand of grass. Insects and mites in turfgrass are adapted to use grasses or grassy habitats for food or development. The biology and management of turfgrass insects has been covered extensively in books (*1–3*) and reviews (*4–6*). These sources comprehensively review turfgrass pest biology and management. This chapter will not be a comprehensive treatment of turf insect biology, but will highlight the key points of plant and pest diagnostics, and management approaches.

Aboveground defoliators

FIGURE 6.1   An egg mass on bermudagrass covered with moth scales.

Defoliation in turfgrass is caused primarily by caterpillars (Lepidoptera) and also a few weevils or billbugs (Coleoptera). Both caterpillars and weevils have chewing mouthparts and are mostly foliage feeders. Turf-feeding caterpillars fall into five groups: cutworms, webworms, armyworms, loopers, and a

Urban Landscape Entomology. https://doi.org/10.1016/B978-0-12-813071-1.00006-3

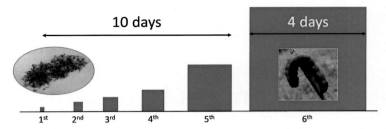

FIGURE 6.2 Relative consumption by fall armyworms over their various life stages. *Adapted from K.L. Flanders, D.M. Ball and P.P. Cobb, Management of Fall Armyworm in Pastures and Hayfields, ANR-1019, 2017, Alabama Cooperative Extension System, Auburn, AL.*

skipper. Turfgrass-infesting caterpillars lay their eggs either as an egg mass on the grass blades (tropical sod webworms, variegated cutworms, and armyworms, (Fig. 6.1), as single eggs usually on the underside of the grass blade (skippers, black cutworms, bronzed cutworms, loopers), or eggs are dropped on the ground by females (sod webworms). The eggs hatch into larvae which develop through a series of larval instars while feeding. Feeding generally happens at night. If during the day, it is generally late in the season (fall) or when conditions are overcast. As larvae increase in size, they consume more and more grass tissue (Fig. 6.2) (*8*). The feeding by smaller larvae can appear as small rounded holes or "windows" in the grass blades. This is why infestations of caterpillars often go unnoticed until the larvae are larger. The caterpillars progress through the larval feeding stages and into a nonfeeding pupal stage. Where they overwinter, it is usually as a pupa or as fully developed larvae (sod webworms) in the soil or the thatch. All species of turf caterpillars can have >1 generation per year and the number of generations and flight period varies depending on the geographic location. A few species like black cutworms and the tropical armyworms (*Spodoptera* spp.) are not able to survive the winter in northern states in the United States (*7,9*). Those insects migrate northward each year from populations in southern states and Mexico. The damage from turf-feeding caterpillars is fairly similar; removal of the grass blades in some instances down to the crown. The grass can also appear yellow or brown, and in need of water (Fig. 6.3). Webbing, caterpillar-produced silk, can be associated with sod webworms and skipper larvae. Burrowing sod webworms (Acrolophinae) live inside a silk-lined burrow in the soil from which they feed. Once they complete development, these silks are often exposed on the surface by lawnmowers. Potter (*1*) compared the appearance of these to empty cigarette wrappers and I have observed large numbers of these silken tubes that appear as litter on a golf course driving range.

Identification of most larval stages of insects is difficult even for trained entomologists. Three characteristics of turf-infesting caterpillars; relative size,

FIGURE 6.3   Damage from fall armyworm feeding in bermudagrass.

stripes or spots, and host grass, are useful to sort caterpillars in turfgrass in to the major groups (Table 6.1). Armyworms are relatively longer (≥3.8 cm [1.5 in.]) than all other grass-feeding caterpillars. Armyworms, cutworms, loopers, and skippers have stripes where sod webworms have spots or blotches. Stripes are lines that run "head to tail" on the caterpillar. The grass loopers (*Mocis* spp.) uniquely have the stripe pattern continue onto the head and have only three abdominal legs (prolegs) compared to five in most turf-infesting larvae. One or more spots or blotches occur on each body segment on webworms but spots never cross body segments like stripes. Finally, the grass that is most commonly damaged separates black cutworms from the others. Black cutworm females prefer creeping bentgrass for egg laying to other grasses (*10,11*) and they rarely outbreak in lawn grasses. Fall

FIGURE 6.4   A fiery skipper larva. The enlarged, helmet-like head is common for skipper larvae.

TABLE 6.1 Quick reference guide to the appearance and relative size of the common groups of grass-feeding larvae.

Common name	Caterpillar appearance	Size when fully grown	Grasses commonly damaged	Occurrence in the United States
Tropical armyworms (*Spodoptera* spp.)				
Fall armyworm	Striped; white wishbone on head; four spots on last abdominal segment	≥3.8 cm (≥1.5 in.)	Any grasses and grass weeds in lawns or on golf courses	Common
Yellow-striped armyworm	Upside down V on head, flattened triangles above the yellow stripe	≥3.8 cm (≥1.5 in.)	Any grass in lawns or on golf courses	Common
Lawn armyworm	Striped, pattern similar to yellow-striped armyworm in large larvae	≥3.8 cm (≥1.5 in.)	Bermudagrass and zoysia grass	Only HI
Other armyworms				
True armyworm	Overwinters in the northern United States; moths and larvae present earlier than tropical species	3.8 cm (1.5 in.)	Any grass in lawns or on golf courses	Common
Tropical webworms (*Herpetogramma* spp.)				
Grass webworm	No stripes on body, only blotches on segments, webbing	2 cm (0.78 in.)	Bermudagrass, Kikuya grass, other warm-season grasses	Only HI
Tropical sod webworm	No stripes on body, only blotches on segments, webbing	2 cm (0.78 in.)	Bermudagrass and St. Augustine grass most common	Southeastern United States

TABLE 6.1 Quick reference guide to the appearance and relative size of the common groups of grass-feeding larvae.—cont'd

Common name	Caterpillar appearance	Size when fully grown	Grasses commonly damaged	Occurrence in the United States
Other webworms (*Crambus* spp., *Acrolophus* spp.)				
Bluegrass webworm, burrowing sod webworm	Blotches on segments, webbing or silk-lined tubes (*Acrolophus* spp.)	2.2—4.4 cm (0.875—1.75 in.)	Wide range of grasses	Common
Grass-feeding loopers (*Mocis* spp.)				
Striped grass looper, striped grassworm	Three pairs of abdominal prolegs; striped pattern on body continues on head	3—5 cm (1.2—2 in.)	Wide range of grasses	Common, but damage is uncommon

armyworms and the other species can feed and develop on most grasses including some weeds. Finally, the fiery skipper (*Hylephila phyleus*), unlike the other caterpillars in turfgrass, appears to have an abnormally large head that looks helmet-like with no contrasting lines (Fig. 6.4).

Aboveground sapsuckers

Sapsuckers that use leaves, stems, and stolons are among the more problematic pests in turfgrass. Because most of these are small, they can easily go undetected in lawns or shipments of sod. The common groups are chinch bugs (*Blissus*), mealybugs, aphids, mites, spittlebugs, and scale insects. Chinch bugs represent a complex of native species in the genus *Blissus* that vary by location in the United States (Table 6.2 (*3*)). They are more commonly a pest of lawns than golf courses. Lawn thatch thickness is associated with greater numbers of chinch bugs in lawns (*12*), yet broad-leaf weeds in lawns are associated with lower hairy chinch bug populations (*13*). Chinch bugs overwinter as adults and

TABLE 6.2 Sap-sucking insects and mites that colonize aboveground parts of turfgrasses.

Common name	Grasses commonly damaged	Occurrence in the United States
Chinch bugs (*Blissus* spp.)		
Southern chinch bug	Most lawn grasses and different grass weeds (goosegrass, pigweed)	Common; southeastern and southwestern states and California
Hairy chinch bug	Cool-season grass lawns	Common; Kentucky and states north into Canada
Western chinch bug (a.k.a buffalograss chinch bug)	Buffalograss and zoysia grass	Central U.S.
Mites (Acari: Tetranychidae and Eriophyidae)		
Tetranychidae: Winter grain, Banks grass, and clover mites	Mostly cool-season grasses except for warm-season grasses for Banks grass mite	Widespread but outbreaks are uncommon
Eriophyidae: Bermudagrass, zoysia grass, buffalograss, St. Augustine grass mites	Mites are specific to their host grasses	Uncommon, but outbreaks can be severe and are anecdotally associated with drought
Mealybugs (Pseudococcidae) and armored (Diaspididae) and soft (Coccidae) scales		
Rhodes grass mealybug	Bermudagrass primarily but many grass hosts	Uncommon, but outbreaks can be severe killing the grass
Buffalograss mealybugs	Buffalograss	Two genera of mealybugs
Bermudagrass mealybug	Bermudagrass	Uncommon, but more likely in Gulf states
Bermudagrass scale	Warm-season grasses	Uncommon armored scale found in most locations where bermudagrass grows
Duplachionaspis divergens	Ornamental *Miscanthus*, zoysia, St. Augustine grass, and bahiagrass	Uncommon armored scale with no common name
Cottony grass scale	Cool-season grass hosts	Uncommon soft scale
Turfgrass scale	Cool-season grasses	Uncommon soft scale

TABLE 6.2 Sap-sucking insects and mites that colonize aboveground parts of turfgrasses.—cont'd

Common name	Grasses commonly damaged	Occurrence in the United States
Aphids (Aphididae)		
Greenbug	Mainly cool-season grasses, a grain pest that can also use turfgrass	Uncommon and minor pest, many generalist aphids can use grasses as hosts
Spittlebugs		
Two-lined spittlebug	Centipede grass, zoysia grass, bermudagrass	Occasional pest, especially in years with above average rainfall

have multiple generations per year. Populations of chinch bugs in damaged lawns are concentrated inside and near areas of damage (*14*). Despite apparent spatial influences on chinch bug populations, no additional research comparing local versus whole lawn management approaches has been published. Adults and immature chinch bugs feed coincidentally, and high populations left unchecked can kill lawns in one season. Chinch bugs insert their mouthparts to access the vascular tissues, yet prefer to feed on phloem (*15,16*). Chinch bug feeding causes grasses to accumulate nonstructural carbohydrates, which are products of photosynthesis commonly used as indicators of stress responses in plants (*17*). There is speculation, but not confirmation, that the saliva (and salivary sheath) used by chinch bugs to move their mouthparts through plants may be toxic or accentuate damage from feeding (*15,16,18*). Chinch bugs have interesting structures in the midgut called crypts that house specific bacteria (mainly *Burkholderia* spp.) (*19*). Southern chinch bugs likely acquire these bacteria by feeding on St. Augustine grass which appears to harbor these bacteria internally (*20*). *Burkholderia* bacteria are important to the ecology of chinch bugs. When the bacteria are experimentally removed, chinch bug immatures have slower development, smaller body size, lower survival, and few can complete development to adults (*19,21*). New technologies such as RNAi may one day yield specific tools that exploit our growing knowledge of chinch bug ecology.

Two families of mites, Tetranychidae (spider mites) and Eriophyidae (gall and stunt mites) can damage grasses (Table 6.2). These mites have multiple generations per year, with a generation usually taking 2 weeks or less. Nymphs and adults co-occur and feed concurrently when present. The ecology of mites in turfgrass is not well known. They tend to outbreak sporadically and can be severe. Without sampling, you will likely never know they are present. Even if detected, few miticides are registered for use in turfgrass which leaves management dependent on less effective, broad spectrum insecticides, and cultural practices. Spider mites insert their modified sucking mouthparts into plants and remove the contents of individual cells. When an infestation is beginning, small stippled spots are obvious on the leaf blade. But high populations will appear as discolored, yellowy patches of turfgrass. Banks grass mite occasionally damage ryegrass overseeded into dormant warm-season grasses. Overseeding provides a green lawn or playing surface during winter. Early feeding damage by mites in overseeded ryegrass will appear similar to drought. Banks grass mite can also outbreak in warm-season grasses and even kill lawns if not managed. Eriophyid mites (*Eriophyes* spp.) in turfgrass have two immature stages then molt to adults. They cause distorted or stunted growth seen as either distorted or straplike foliage or a shortening or compressing of internodes. Internodes are the length of stem between leaves and when shortened the foliage appears bunched together.

Mealybugs (Pseudococcidae) and armored (Diaspididae) and soft (Coccidae) scale insects are uncommon, but occasionally serious pests in turfgrass. As noted with mites, detection and outbreaks of scale insects and mealybug are uncommon so there are a limited number of publications on the biology, ecology, and management. Most have multiple generations per year, especially in the warmer climatic zones in the United States. They have sedentary habits and only spread naturally through immature stages called crawlers. Humans may spread these pests by moving infested sod or plugs. Crawler movement on maintenance equipment, birds, mammals, or other insects in the landscape are possible (*22*), although not documented.

Mealybugs and soft scales in turfgrass are conspicuous (Fig. 6.5), and feeding can produce honeydew and sooty mold and recruit ants. In the

FIGURE 6.5   Rhodes grass mealybugs in turfgrass. The image on the left shows the anal filament with a droplet of honeydew. Growth of sooty mold fungi is evident around the mealybug.

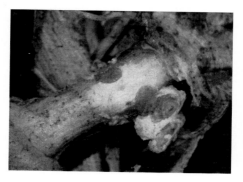

FIGURE 6.6   Armored scale insects (Diaspididae) are cryptic and often go unnoticed in turfgrass. Few are well studied and they are assumed to contribute to reduced vigor in infested grasses.

southeastern and southwestern United States, invasive Argentine and imported fire ants tend and carry mealybugs which can facilitate outbreaks (23). The ants benefit by having a source of free sugars provided by the phloem-feeding aphids or soft scale insects. Armored scales are more cryptic in appearance and less apparent in grass samples (Fig. 6.6). They do not produce sugar-rich honeydew, are not tended by ants, and populations generally grow slower than mealybugs and soft scales. The detection of an exotic armored scale insect, *Duplachionaspis divergens* (24), suggests that there are scale insects or mealybugs in turfgrass, but there are likely many species that go undetected.

The two-lined spittlebug is a grass-feeding pest species but it is not a specialist on grasses. It has two generations per year in most places in the United States where it occurs. Overwintered eggs hatch in spring and one or more nymphs develop inside a frothy mass (Fig. 6.7) that gives spittlebugs their common name. Spittlebug adults and immatures feed on plant xylem. The immature stages consume large amounts of xylem fluid and expel the froth for protection from predators (25). The adult stages feed on certain hollies or turfgrass. Dahoon (*Ilex cassine*) and American (*Ilex opaca*) hollies near turfgrass may increase damage by spittlebugs (26). Damage to turfgrass will appear as either wilted or brown grass and an overall dry appearance, despite excessive rainfall or adequate irrigation. In fact, there are greater populations of spittlebugs under wetter conditions (27). Adults inject a toxin during feeding which can cause streaking or death to the tops of grasses (28). Centipede grass, bermudagrass, and zoysia grass lawns are more likely to host two-lined spittlebugs.

Identification of aboveground sap-sucking insects is easy for chinch bugs and two-lined spittlebugs, but confirmation of mites, scales, mealybugs, and aphids requires slide mounting, a microscope or hand lens, or submitting a sample to a laboratory. Chinch bug adults will have wings but there are short-winged and long-winged adults. This can be misleading because most insects

FIGURE 6.7   Spittle mass containing the nymph of the two-lined spittlebug. The nymphs produce the spittle mass as a way to use excess water consumed from xylem feeding.

lacking fully developed wings are still immature (nymphs). The abundance of either adult type can vary through the season. Adult chinch bugs have a dark spot in the wings and the head will be smaller than the width of the body. Nymphs have a light-colored band across the back where the wings are located in adults. The first two stages have an orange-red color body and head color compared to a dark-colored body and head in older nymphs and adults. Since these first two immature stages differ in coloration from the other stages, they can be overlooked as being chinch bugs. Two-lined spittlebug adults have two distinct reddish orange lines on the wings that contrast well with the darker body. Adults are most likely seen hopping away from lawnmowers. My students taught me that the spittle masses are thought, by some, to be snake spit. The spittle mass is produced by the immatures in response to feeding and does not remain after they have reached the adult stage or died. Therefore, a spittle mass on turfgrass essentially is a positive identification for actively feeding two-lined spittlebug nymphs.

The other sap-sucking insects and mites found above ground can be identified easily to order and perhaps family, but will likely need to be confirmed by a trained entomologist. Spider mites and eriophyid mites have different body forms. Most mites have the same form as ticks: oval, with eight legs. Spider mites are about the size of a period of text, and you can see them without a microscope. Eriophyid mites in turfgrass are only visible using a microscopic or hand lens. Unlike spider mites they have elongated bodies and only four legs. Soft scale insects and mealybug can be mistaken for noninsects because they often do not have external appendages (legs, antennae) that help with identification. Mealybugs will usually be covered with a white flocculence. Flocculence is a white coating (flock) like the coat on sheep. Many of the adult females will be enlarged because of their ovisac. Rhodesgrass mealybug, for example, has a flocculent female with a large ovisac, and they have a distinct tube, called an anal filament, from which they drop their honeydew (Fig. 6.5). Scale insects in grass are cryptic and difficult to recognize as insects. An insect identifier is needed to get confirmation of the genus or species. In the United States, your state insect identifier is part of the National Plant Diagnostic Network (https://www.npdn.org/home) and the location of your state's diagnostic laboratory can be found at that website. Since scale insects, mealybug, and aphids are more common on ornamental plants, there will be more discussion about them in Chapter 7.

Turfgrass insects found in soil

Not all insects that live underneath turfgrass in the soils or thatch layer will be root feeders. Ants, for example, are predatory insects that do not feed on grass, but their colonies are associated with soil and usually the soil under turfgrass. White grubs, weevils/billbugs, crane flies, mole crickets, and ground pearls are the main groups of grass-feeding insects found in soil. Worms, earthworms (Phylum: Annelida) and nematodes, can also be pests and occur in soils under turfgrass. You will rarely see a nematode, flatworm, or fluke because of their size, but earthworms are common in turfgrass especially in the spring and fall. They do not eat turfgrass, but deposit castings on the grass surface. On golf courses, castings interfere with ball roll, but in lawns they can smother entire sections of turfgrass (Fig. 6.8). Most of the casting damage in turfgrass is caused by exotic earthworms (*29*). These castings can also cycle nematodes from below ground to the turf surface (*unpublished data*). Pests in soil can damage turfgrass by either mounding soil on top of grass, by feeding on roots, or by mechanical damage (tunneling through soil and roots). Mole crickets, for example, may cause more than one type of damage, mounding and root feeding.

White grubs (Scarabaeidae) are larvae of scarab beetles found worldwide in virtually all grass habitats. They are represented by about 20 species in 10 genera in turfgrass in the United States with native masked chafers (*Cyclocephala* spp.) (*30*) and May/June beetles (*Phyllophaga* spp.) (*31,32*) being

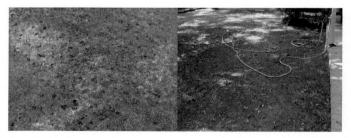

FIGURE 6.8 Earthworms in turfgrass, particularly exotic species, can cause problems through excessive castings. On golf course tees and greens (left), casts interfere with ball roll. In home lawns (right), earthworms can cast so heavily that they smother large sections of grass.

most widespread. The number of species varies regionally with the most diversity of species occurring in the northeastern United States (see Ref. (*1*) for distribution maps of species in the United States). Most species of white grubs have one generation per year (i.e., annual white grubs). However, many species in the genus *Phyllophaga* take >1 year to complete development. In warmer climates (e.g., Florida and California), scarab species with one generation elsewhere may have a partial or complete second generation (*33*). Grubs are present most of the year (9 months; September to April) in the soil as larvae or pupae. The adult stages of some species like Japanese beetles are pests of trees and shrubs. For others, the adults are drab-colored beetles that are mostly night active and therefore not commonly seen. Most white grubs mainly feed on either organic matter or grass roots. Some, like Green June beetle larvae or sugarcane beetle adults, do mechanical damage. As grubs consume plant material they are inevitably consuming soil and soil bacteria (*34*). Sugarcane beetles are one of the few scarabs that damage turfgrass in the adult stage. Damage from this beetle is often associated with lights. Street lamps or security lights adjacent to turfgrass recruit and accumulate the adults. Tunneling and likely feeding below the surface causes damage (*35*).

Grass-feeding billbugs (*Sphenophorus* spp.) and the annual bluegrass weevil (*Listronotus maculicollis*) are snouted beetles in the family (Curculionidae). The adults have chewing mouthparts but have an elongated rostrum that looks beaklike. Within the United States, the annual bluegrass weevil is a pest of golf courses in the northeastern states (*36*) but billbug species are found nationwide (*37*). There are four major species of native billbugs in the United States. In the eastern states, hunting billbug (*Sphenophorus venatus*) is the most prevalent either in traps or associated with damage in turfgrass (*38–41*). The number of generations varies widely by species and location. Annual bluegrass weevil has two generations per year (*36*) and hunting billbug has one generation in northern states and eight to nine generations in Florida (*38*). Where hunting billbugs have more than one generation per year, the generations overlap producing populations with multiple life stages present at the same time (*38,39*).

Overwinter, adults are common where the sidewalk and grass meet (*37*). They may also overwinter as larvae in the soil. Annual bluegrass weevils overwinter as adults in the leaf litter of trees adjacent to the golf course (*42*). The biology of grass-feeding weevils and billbugs is similar. Adults do not commonly fly but walk along the grass surface. In spring, females feed on grasses and create notches where they lay a few eggs in each notch. The eggs hatch and the weevils feed inside the grass plant for their first one to two instars. Basically, when the grass plant is no longer large enough to contain the growing larva, it enters the soil and continues to develop as a root feeder. They pupate in the soil and then the adults emerge (*36–38,40*). The larvae inside the stem are only detectable by dissection or with the salt disclosing solution detailed in Chapter 5.

FIGURE 6.9 Raster patterns are used to identify white grubs to genus or species. This raster pattern is for Asiatic garden beetle.

Species of white grubs overlap in their geographic distributions but there are still often one or two dominant species in an area. For this reason, most control recommendations are based on data for these common species. Where mixed populations occur, it is possible that insecticides or microbial insecticides provide better control of some species relative to others (*43,44*). For this reason, it is sometimes necessary to identify grubs to genus or species. This is possible in the field by observing the pattern of hairs on their raster (Fig. 6.9). Furthermore, white grubs often co-occur in soil with other beetle larvae. White grubs are easily separated from grass-feeding weevils and billbugs by the presence of three pairs of thoracic legs just behind the head. Billbug larvae are always legless but have a head that contrasts with the body (Fig. 6.10). Predatory beetle larvae in soil will have mouthparts (mandibles) that project straight out from the front of the head. The mouthparts on white grubs and billbug larvae point downward. Unlike white grubs, billbug larvae can only be identified to family in the field. As a rule of thumb, a legless grub collected from turfgrass will likely be a billbug unless collected under an oak or pecan tree. In those uncommon situations, the larva in the soil could be a billbug or one of the nut-feeding weevils that complete development in the soil.

Mole crickets are not as widespread as white grubs but equally or more damaging where they do occur. Mole crickets that damage turfgrass in the United States are either the native species, *Neocurtilla hexadactyla*, or three introduced species in the genus *Neoscapteriscus* (formerly *Scapteriscus*). The tawny (*Neoscapteriscus vicinus*) and southern (*Neoscapteriscus borellii*) mole crickets occur across the Gulf Coast states and north in the coastal areas of the Carolinas. Populations of the third species, the short-winged mole cricket (*Neoscapteriscus abbreviatus*), occur only in Florida (*45,46*). Tawny and southern mole crickets are more widespread and damaging. Related *Neoscapteriscus* spp. are what is typically referenced as "mole crickets" either

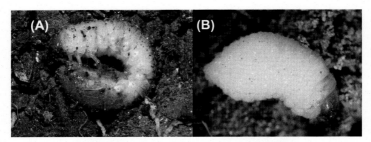

FIGURE 6.10 White grubs (Scarabaeidae) (A) are easily separated from larvae of weevils and billbugs (B) by the presence of three pairs of thoracic legs.

online or in Extension publications. It is common for multiple species of mole crickets to be present in a lawn or golf course, but usually one species will be more abundant. Native and introduced species of *Gryllotalpa* and *Neocurtilla* may also occur in turfgrass, but populations that cause damage are uncommon (*4*). Species of *Gryllotalpa* and *Neocurtilla* have a broader geographic range with detections through the midwestern and northeastern states. Tawny and southern mole crickets take 1 year to develop in the United States except in southern Florida where southern mole crickets have a second generation. The biologies of other species are not thoroughly studied, but presumed to have either one generation per year in warmer climates, or take 2 years to develop in northern states (*46,47*).

Mole crickets are omnivores, consuming insects or earthworms, or feeding on plant roots or tender shoots. Yet, tawny mole crickets prefer plants as food, and southern mole crickets prefer animal tissue (*48,49*). Adults and immatures live and develop in elaborate tunnels with architecture unique to each species. Studies (*50,51*) have used CT and X-ray images of tunnels developed under laboratory conditions and others have used materials to make casts of tunnels (*50,52*). Tunnels are more extensive in sandy soils than in clay soils (*50*) which help explain why mole crickets are more problematic in the sandier soils of coastal areas or where playing surfaces are amended with sand. Adults emerge from these tunnels 1 or 2 times per year for mating and dispersal flights (usually Feb–May annually). Not long after dusk, males in the underground tunnels use species-specific songs (like field crickets) to call females (*53*). After mating, females lay egg clutches in egg chambers underground. The egg laying period for mole crickets can last for 60 or more days producing the next generation in May to August. Egg hatch is important for management (*46*). The time needed for eggs to hatch varies with soil temperature and species of mole cricket. Hayslip (*54*) provides temperatures for egg laying and incubation of mole crickets, but landscape managers and golf course superintendents use the soap disclosing solution discussed in Chapter 5 to detect newly hatched nymphs (Fig. 6.11). Nymphs develop through the late summer and fall in soil

FIGURE 6.11   Mole cricket egg clutch (top left) and newly hatched nymph (lower left). The flowering perennial *Agapanthus* is an indicator of when to check turfgrass for egg hatch *(49)*.

and can overwinter as either nymphs or adults *(46)*. Therefore, insects that hatch in late spring and summer of the first year will be present through winter and into early spring of the next year *(47)*.

Pest mole crickets are in one insect family (Gryllotalpidae),recognized by their enlarged digging forelegs. Separating individuals to genus or even species requires looking at claw on the forelegs or coloration on the pronotum *(46,47)*. *Neoscapteriscus* mole crickets have a two-clawed dactyl on their foreleg compared to a four-clawed dactyl on native species *(55)*. Furthermore, tawny mole crickets have a V-shaped space compared to a U-shaped space for southern mole crickets *(46)*. To recall this, just remember there is a "U" in southern but not tawny. Species identification of pest mole crickets is somewhat academic because the published control data suggest that tawny and southern are similarly susceptible to soil applied insecticides. However, the species have different consumption rates of insecticide bait formulations that may affect control with these baits *(56)*.

Ground pearls (Hemiptera: Sternorrhyncha: Margarodidae) are primitive scale insects that feed on grasses and sugarcane. They are biologically unusual as a sap-sucking insect that feeds below ground on roots. Ten genera and 8000 species occur worldwide with species in seven genera reported from grass hosts. Only two of those seven genera occur in North America, but the others may be pests of turfgrass where they occur *(6,57)*. In North America, *Margarodes meridionalis* and *Eumargarodes laingi* are the ground pearls reported from turfgrass (*58,59*). The occurrence of ground pearls is sporadic and the life cycle is poorly understood. They have one generation per year and the cyst stage is the longest and most commonly detected life stage in turfgrass (*2,57,59*). The common name comes from the rounded, pearl-like cyst stage. A cyst lacks legs and external appendages and superficially resembles a slow-release fertilizer pellet in soil. Inside the cyst, one second instar nymph feeds by their sucking mouthparts on grass roots (*57,59,60*) and develops until the female emerges (*57*). The female appears as a pinkish-red color and somewhat pear-shaped insect, with legs and antennae (Fig. 6.12). Males are rare in ground pearls (*57*). Unlike other scale insects that have sedentary females, female ground pearls disperse. Once a suitable location is found, the female constructs an egg chamber, deposits about 100 eggs, and then dies (*57,59*). Once they hatch, the first instar nymph, which also has well-developed legs and antennae, will disperse then attach to roots. Once attached, they feed for about 2 months before molting to a second instar. The second instar remains at that location and forms the cysts (*57*). Cysts are aggregated in the soil (*59*), and populations of ground pearls are excessive (cyst counts) when turfgrass is severely damaged. For example, damaged turfgrass may have more than 4000 cysts per pint of soil (*58,61*). There are no thresholds for number of cysts in relation to damage in turfgrass, but thresholds for ground pearls exist in sugarcane (*60*).

Ground pearls prefer sandy soils in dryland conditions (*57*), and can cause severe damage or even kill stands of centipede grass, or Tifgreen and Tifdwarf varieties of bermudagrass (*59,62*). Feeding by ground pearls produces an unhealthy, yellow turf usually in patches. The appearance worsens, during drought the stand often turns brown and eventually dies. When diagnosing ground pearls, people may use phrases like "grass never grows there" or "we resodded that area multiple times but the grass just dies." These are not unique symptoms of ground pearls but this should prompt a soil sample, or looking for females in late April to August (*63*).

FIGURE 6.12   Female ground pearl (Margarodidae).

Another odd pest of turfgrass is the cranberry girdler (*Chrysoteuchia topiaria*). Taxonomically, this is a type of sod webworms but larvae of the cranberry girdler attack the roots and crown of cool-season grasses. Unlike other sod webworms that are generally grass specialists, the cranberry girdler damages grass as well as woody plants. The biology of this species has been most studied in cranberries, forestry, and grass seed crops. In turfgrass, the biology is based on light trap studies and details of outbreaks recorded by Extension specialists (*64*). In turfgrass, problems are restricted to northern states and occasional but severe outbreaks in lawns or golf courses. There is one generation per year with the larval stage feeding in the summer following egg hatch, then overwintering. The larvae typically do not feed in the following spring before pupating. Cranberry girdler larvae collected from cool-season grasses would have the same characteristics previously mentioned for other sod webworms in turfgrass. If the larvae are confirmed as cranberry girdler, the control measures may require posttreatment irrigation or rainfall to move the insecticide into crown or root zone (*65*).

Several species of crane flies (Tipulidae: *Tipula* spp.) damage pasture and turfgrass as larvae (*6,66*). Of those, two introduced species, European crane fly (*Tipula paludosa*) and the common crane fly (*Tipula oleracea*) are significant pests of cool-season turfgrass in the Pacific northwest and northeastern United States (*67*) and in Canada (*66,68*). There are similarities in the development of crane flies, but European crane fly has one generation per year and common crane flies have two generations per year. Adults are large flies, sometimes called mosquito hawks because they are incorrectly believed to attack adult mosquitos. The adults may not feed or may feed on nectar. Adults are active just after sunset and can occur in large numbers. Eggs are deposited in or around moist turfgrass. Larvae called "leatherjackets" are gray to brown colored, cylindrical, tapering slightly at both ends. They develop in soil through four instars usually in the upper 2.5 cm (1 in.) of soil but move within the soil profile depending on soil moisture (*68,69*). Leatherjackets are 2.5−3.5 cm (1−1.5 in.) long just before they pupate (*68*). The pupae protrude from the grass slightly before the adults emerge. European crane fly adults emerge in fall and eggs hatch in September. Larvae develop through two instars in 2 months. Third instars overwinter, then become active in April. Fourth instar larvae are found April through August. During this period, larvae may stop feeding and developing, and enter a resting phase called aestivation (summer dormancy). Common crane flies have a spring and fall emergence with adult emergence occurring earlier in spring in the Pacific Northwest compared to the Northeastern United States (*66*).

All cool-season grasses as well as white clover are food for larvae (*69,70*). Leatherjackets feed on the surface on blades, stems, and crowns at night. During the day they feed below the surface on root hairs, roots, and thatch (*6,66*). They can often be observed or collected at night or the early morning as they wander or migrate in search of food (*68*). In New York, bentgrass putting greens can be severely damaged by European crane fly (*71*). Fourth instar larvae of the European crane fly active at night on putting greens can be mistaken for black cutworms. In the Pacific northwest, both species damage lawns, but populations have decreased and it is no longer considered a serious annual pest (*68*). In New York state, however, high densities of larvae and adults are found on golf courses, especially in creeping bentgrass (*71*). Poorly maintained lawns are particularly susceptible to lower densities of leatherjackets. Well-maintained lawns can tolerate 40–60 larvae per 0.1 m^2 (per ft^2) compared to only 12–15 larvae per 0.1 m^2 (per ft^2) in lawns under traffic or shade stress. Bird and vertebrate animals will search infested turf for leatherjackets. This may be a good first indication that leatherjackets are present (*66*). Sampling for larvae is done by collecting four plugs, 10–15 cm (4–6 in.) wide, of turf at least 5 cm (2 in.) deep with a spade, golf course cup cutter, or a large knife (*66,68*). The samples should then be pulled apart and shaken over a tray as discussed in Chapter 5. Larvae of crane flies can be only separated to genera using morphological characters (*72*). The larvae or pupae can be collected from turfgrass at night and held for adults to emerge. Otherwise, identification of larvae or pupae to species requires assays using insect protein or DNA (*69,73*). Adult crane flies are large and the antennal, wing, and eye characters needed to separate them are not difficult to observe. Not all crane flies observed will be pests so adult identification is an important first step in assessing concerns about crane flies.

Beneficial arthropods in turfgrass

Ants are abundant in turfgrass. Their roles as natural enemies of turfgrass pests will be discussed later, but this section will focus on mound building and human interactions. Ants are often the most collected insects when operating traps in turfgrass (*74–77*). The three most documented ants in lawns and on golf courses are *Lasius neoniger* (oddly called the turfgrass ant), *Solenopsis molesta*, and *Solenopsis invicta* (imported fire ants) (*74–77*). These species are mound-building ants. Mounds are the surface indication of an otherwise hidden, subterranean ant colony. Mounds are made of the underlying soil. *L. neoniger* creates small mounds, 10–60 mm (0.4–2.4 in.) wide and just 15–20 mm (0.6–0.8 in.) tall, with a hole in the apex where ants enter and exit the mound. In lawn grass or golf course roughs, these mounds would never appear above the mowing height of most grasses. However, they are apparent and can interfere with play on shorter mown putting greens and tees (*77*). Imported fire ant mounds in turfgrass are taller: 84–181 mm (≥3 in.) tall in sandy soils (*78*) and >300 mm (12 in.) in clay soils with no central hole in the apex. Imported fire ants enter and exit the colony below ground, not through the top of the mound. Fire ant mounds are well above the height of most maintained turfgrass meaning the top of the mound will be cut during mowing. Mowing over mounds can dull mower blades, leading to more frequent maintenance on mowers. Mowing over fire ant mounds can also cause a swarm response and defensive stings. Rhoades et al. (79) estimated about 31% of people within the imported fire ant distribution in the United States (9.3 million total) may be stung annually. Fatalities occur annually and can result from less than five stings in an allergic individual (79). Many other invasive ants including crazy ants (*Nylanderia fulva*) and Argentine ants (*Linepithema humile*) are also abundant in suburban areas in the southern United States but neither of those species sting or produce large mounds. Finally, imported fire ants uniquely impact quality of life and home economics. Medical care costs, control costs, and repairs due to imported fire ant mounding inside of air-conditioning units and electrical and cable boxes can be costly for urban residents and golf courses. In a survey of the five major metroplexes in Texas, Lard et al. (80) looked at the economic impacts on suburban residents, golf courses, and schools. Repair and replacement costs on air-conditioning units due to fire ants ranged from $6 to $83 annually per household. In the same study, golf courses reported about $100,000 annually spent on control and equipment replacement costs due to imported fire ants. About 50%–60% of that total was used to replace irrigation and circuit boxes damaged by fire ant habitation (80).

All turfgrass pests are subject to attack by arthropod natural enemies, bird or animal predation, and insect pathogens. The occurrence of these biological agents is well documented under field conditions and believed to buffer turf-grass from outbreaks of pests (81). The most common groups of arthropod natural enemies are ants, many types of predatory beetles, and spiders. Mound-building ants, discussed previously, have dual roles as pests and beneficials. Imported fire ants, *Solenopsis* spp. in general and *L. neoniger* are effective predators mainly of insect eggs in turfgrass (75,76,82–84). Most of what we know about the occurrence of arthropod predators in turfgrass come from studies where the diversity and abundance were monitored before and after an insecticide application (e.g., 75,84). Another set of studies examined the potential of certain predators to consume turfgrass insects under field or laboratory conditions (25,75–76,82–87). For example, ants can consume about 70% of insect eggs placed in turfgrass for <24 h (76,82,83). Most of the predatory insects and spiders in turfgrass are active at night (86) and are rarely observed during the day. In additional to beetles and spiders, there are families of predatory insects (true bugs) with sucking mouthparts. Big-eyed bugs (Family: Geocoridae), minute pirate bugs (Family: Anthocoridae), and assassin bugs (Family: Reduviidae) are among the most common. Big-eyed bugs are important predators of chinch bugs and fortunately very abundant in grass species and cultivars prone to chinch bug outbreaks (St. Augustine grass, buffalograss, and zoysia grass (87,88). However, the casual observer may assume that insects with sucking mouthparts in a susceptible turfgrass are chinch bugs. Predatory big-eyed bugs get the common name because the head is wider than the body. This difference is easily seen with an inexpensive hand lens. When I was an Extension Specialist, one landscape manager submitted a turf sample claiming a chinch bug outbreak. Upon examination, the samples contained mostly big-eyed bugs. Without careful observation, that lawn may have been sprayed to control beneficial insects.

Paper wasps (*Polistes* spp.) are caterpillar predators that hunt armyworms and other caterpillars in turfgrass during the day. Their presence on turfgrass should prompt sampling for armyworms or other larvae (*89*). Other wasps in turfgrass are parasitoids: insects that lay eggs on or inside other insects that are later killed by the developing larva (this act is called parasitism). Parasitoid wasps of white grubs, chinch bugs, mole crickets, armyworms, black cutworms, and Rhodesgrass mealybugs are either naturally occurring or have been introduced as part of biological control programs. Wasps that attack grubs and mole crickets must locate the patch of grass infested by grubs likely using volatiles (*90*), then search for their hosts underground in the dark following cues like insect body odor or frass (*91*). The female then lays an egg on the host and exits the soil. Other parasitoid wasps may insert their eggs into caterpillar eggs. The caterpillar then develops with the parasitoid eggs inside. There are a few parasitoids of white grubs, fire ants, and mole crickets that are flies. Those fly parasitoids deposit an egg on the outside of the insect, primarily the adult stage, and then the developing larvae kills the insect (*45,92,93*). A few predatory insects, but no parasitoids, are commercially available for purchase and release in turfgrass. Papers documenting the release of predatory insects in turfgrass are not available, perhaps because so many of the generalist predators already exist in the system.

The actions of insect natural enemies can be enhanced or hindered by the local environment. Parasitoid wasps and flies often get nectar from flowering plants and planting of flowering plants adjacent to turfgrass can positively impact the activity of natural enemies. For example, strips planted with alyssum, coreopsis, and switchgrass alongside a fairway locally increase predators and parasitoids which can increase predation of black cutworm larvae in fairways adjacent to the strip (*85*). Flowering peonies provide sugar through extrafloral nectaries to *Tiphia* wasps that parasitize white grubs. However, to increase parasitism the peonies must be within 1 m of the grub infestation (*94*). *Tiphia* wasps may also be better at finding grubs in different grasses. Masked chafer grubs in *Zoysia japonica* have greater parasitism by *Tiphia* wasps. Similarly, creeping bentgrass was the lowest of all grasses for parasitized Japanese beetle grubs (*95*). There are several studies linking the level of resistance in grasses to pests to the success or failure of natural enemies. Fungal endophytes in cool-season perennial ryegrass produce toxic alkaloids and negatively affect some turfgrass pests. The same alkaloids can also hinder the development of some parasitoids when caterpillars are feeding on endophytic grasses (*92*). Similarly, predatory true bugs have a greater impact on fall armyworms feeding on grasses with intermediate resistance

FIGURE 6.13 Animals—birds, skunks, moles, and armadillos—can damage turfgrass while searching/foraging for insects in turfgrass. Image A shows numerous small pokes that the animal(s) used when searching. Image B shows a hole, approximately 30 cm (12 in.) deep produced by an armadillo likely search for mole crickets. Both figures show a relatively healthy and green turfgrass stand damaged by the animals but not turfgrass pests.

compared to grasses that are either more susceptible or resistant to fall armyworms (*96*).

Finally, animal predation on soil-dwelling insects by skunks, moles, armadillos, and racoons (*1,95*) can be significant and often more damaging than the insects. Skunks mostly damage to cool-season grasses whereas armadillos are the primary varmint damaging warm-season grasses (Fig. 6.13). Moles feed on earthworms and soil insects and activity is often in response to earthworm or insect activity. In most situations, the varmints are able to detect earthworms or soil insects at levels that may not cause damage. Some turfgrass managers will use insecticides as curative or rescue treatments to address customer complaints about foraging skunks or armadillos. Insecticides likely eliminate the food source and force foraging animals to go elsewhere. Volatile organic fertilizers applied to the turf surface and not irrigated may discourage foraging in cool-season grasses by skunks for 3—4 weeks after application (*97*).

Highlights

- Turfgrass hosts a community of herbivores adapted to consume grass and arthropod natural enemies and pathogens associated with these herbivores.
- Despite the prevalence of turfgrass in urban environments, the ecologies are well understood for just a few of the most severe pests.
- Damage to turfgrass can result from plant feeding but insects, earthworms, or even animals dig or mound soil on the surface also causing damage.
- Turfgrass pest management is currently dependent on insecticides but understanding pest ecology and biological controls may provide alternatives to insecticides.

References

1. Potter, D. A. *Destructive Turfgrass Insects: Biology, Diagnosis, and Control;* Ann Arbor Press: Michigan, 1998.

2. Vittum, P. J.; Villani, M. G.; Tashiro, H. *Turfgrass Insects of the United States and Canada,* 2nd ed.; Comstock: Wisconsin, 1999.

3. Brandenburg, R. L.; Freeman, C. P. *Handbook of Turfgrass Insects,* 2nd ed.; Entomological Society of America: Maryland, 2012.

4. Held, D. W.; Potter, D. A. Prospects for Managing Turfgrass Pests With Reduced Chemical Inputs. *Annual Review of Entomology* **2012**, *57,* 329−354.

5. Koppenhöfer, A. M.; Latin, R.; McGraw, B. A.; Brosnan, J. T.; Crow, W. T. Integrated Pest Management. In *Turfgrass: Biology, Use and Management, Agronomy Monograph,* 56, Horgan, B. P., Stier, J. C., Bonos, S. A., Eds.; American Society of Agronomy: Madison, Wisconsin, 2013; pp 933−1006.

6. Williamson, R. C.; Held, D. W.; Brandenburg, R.; Baxendale, F. Turfgrass Insect Pests. In *Turfgrass: Biology, Use and Management, Agronomy Monograph,* 56, Horgan, B. P., Stier, J. C., Bonos, S. A., Eds.; American Society of Agronomy: Madison, Wisconsin, 2013; pp 809−890.

7. Nagoshi, R. N.; Meagher, R. L.; Hay-Roe, M. Inferring the Annual Migration Patterns of Fall Armyworm (Lepidoptera: Noctuidae) in the United States from Mitochondrial Haplotypes. *Ecology and Evolution* **2012**, *2* (7), 1458−1467.

8. Flanders, K. L.; Ball, D. M.; Cobb, P. P. *Management of Fall Armyworm in Pastures and Hayfields, ANR-1019;* Alabama Cooperative Extension System: Auburn, AL, 2017; p 8.

9. Showers, W. B. Migratory Ecology of the Black Cutworm. *Annual Review of Entomology* **1997**, *42,* 393−425.

10. Williamson, R. C.; Potter, D. A. Oviposition of Black Cutworm (Lepidoptera: Noctuidae) on Creeping Bentgrass Putting Greens and Removal of Eggs by Mowing. *Journal of Economic Entomology* **1997**, *90,* 590−594.

11. Hong, S. C.; Obear, G. R.; Liesch, P. J.; Held, D. W.; Williamson, R. C. Suitability of Creeping Bentgrass and Bermudagrass Cultivars for Black Cutworms and Fall Armyworms (Lepidoptera: Noctuidae). *Journal of Economic Entomology* **2015**, *108,* 1954−1960.

12. Kortier Davis, M. G.; Smitley, D. R. Relationship of Hairy Chinch Bug (Hemiptera: Lygaeidae) Presence and Abundance to Parameters of the Turf Environment. *Journal of Economic Entomology* **1990**, *83,* 2375−2379.

13. Majeau, G. J.; Brodeur, J.; Carrière, Y. Lawn Parameters Influencing Abundance and Distribution of the Hairy Chinch Bug (Hemiptera: Lygaeidae). *Journal of Economic Entomology* **2000**, *93,* 368−373.

14. Cherry, R. Spatial Distribution in Southern Chinch Bugs (Hemiptera: Lygaeidae) in St. Augustinegrass. *Florida Entomologist* **2001**, *84,* 151−153.

15. Painter, R. H. Notes on the Injury to Plant Cells by Chinch Bug Feeding. *Annals of the Entomological Society of America* **1928**, *21,* 232−242.

16. Backus, E. A.; Rangasamy, M.; Stamm, M.; McAuslane, H. J.; Cherry, R. Waveform Library for Chinch Bugs (Hemiptera: Heteroptera: Blissidae): Characterization of Electrical Penetration Graph Waveforms at Multiple Input Impedances. *Annals of the Entomological Society of America* **2013**, *106,* 524−539.

17. Heng-Moss, T.; Macedo, T.; Franzen, L.; Baxendale, F.; Higley, L.; Sarath, G. Physiological Responses of Resistant and Susceptible Buffalograss to *Blissus occiduus* (Hemiptera: Blissidae) Feeding. *Journal of Economic Entomology* **2006**, *99,* 222−228.

18. Ramm, C.; Wayadande, A.; Baird, L.; Nandakumar, R.; Madayiputhiya, N.; Amundsen, K.; Donze-Reiner, T.; Baxendale, F.; Sarath, G.; Heng-Moss, T. Morphology and Proteome Characterization of the Salivary Glands of the Western Chinch Bug (Hemiptera: Blissidae). *Journal of Economic Entomology* **2015**, *108*, 2055–2064.

19. Xu, Y.; Buss, E. A.; Boucias, D. G. Impacts of Antibiotic and Bacteriophage Treatments on the Gut-Symbiont-Associated *Blissus insularis* (Hemiptera: Blissidae). *Insects* **2016**, *7*, 61.

20. Xu, Y.; Buss, E. A.; Boucias, D. G. Environmental Transmission of the Gut Symbiont *Burkholderia* to Phloem-Feeding *Blissus insularis*. *PLoS One* **2016**, *11* (8), e0161699.

21. Boucias, D. G.; Garcia-Maruniak, A.; Cherry, R.; Lu, H.; Maruniak, J. E.; Lietze, V. U. Detection and Characterization of Bacterial Symbionts in the Heteropteran, *Blissus insularis*. *FEMS Microbiology Ecology* **2012**, *82*, 629–641.

22. Magsig-Castillo, J.; Morse, J. G.; Walker, G. P.; Bi, L. B.; Rugman-Jones, P. F.; Stouthamer, R. Phoretic Dispersal of Armored Scale Crawlers (Hemiptera: Diaspididae). *Journal of Economic Entomology* **2010**, *103* (4), 1172–1179.

23. Helms, K. R.; Vinson, S. B. Apparent Facilitation of an Invasive Mealybug by an Invasive Ant. *Insectes Sociaux* **2003**, *50*, 403–404.

24. Evans, G. A.; Hodges, G. S. *Duplachionaspis divergens* (Hemiptera: Diaspididae), a New Exotic Pest of Sugarcane and Other Grasses in Florida. *Florida Entomologist* **2007**, *90*, 392–393.

25. Nachappa, P.; Guillebeau, L. P.; Braman, S. K.; All, J. N. Susceptibility of Twolined Spittlebug (Hemiptera: Cercopidae) Life Stages to Entomophagous Arthropods in Turfgrass. *Journal of Economic Entomology* **2006**, *99* (5), 1711–1716.

26. Braman, S. K.; Ruter, J. M. Preference of Twolined Spittlbug for *Ilex* Species, Hybrids, and Cultivars. *Journal of Environmental Horticulture* **1997**, *15*, 211–214.

27. Braman, S. K.; Abraham, C. M. Twolined Spittlebug. In *Handbook of Turfgrass Insects;* Brandenburg, R. L., Freeman, C. P., Eds., 2nd ed.; Entomological Society of America: Maryland, 2012.

28. Byers, R. A.; Wells, H. D. Phytotoxemia of Coastal Bermudagrass Caused by the Two-Lined Spittlebug, *Prosapia bicincta* (Homoptera: Cercopidae). *Annals of the Entomological Society of America* **1966**, *59*, 1067–1071.

29. Redmond, C. T.; Saeed, A.; Potter, D. A. Seasonal Biology of the Invasive Green Stinkworm *Amynthas hupeiensis* and Control of its Casts on Golf Putting Greens. *Crop Forage and Turfgrass Management* **2016**, *2*. https://doi.org/10.2134/cftm2016.0006.

30. Gyawaly, S.; Koppenhöfer, A. M.; Wu, S.; Kuhar, T. P. Biology, Ecology, and Management of Masked Chafer (Coleoptera: Scarabaeidae) Grubs in Turfgrass. *Journal of Integrated Pest Management* **2016**, *7*, 1–11.

31. Luginbill, P., Sr.; Painter, H. R. *May Beetles of the United States and Canada, Tech. Bull. No. 1060;* US Department of Agriculture: Washington, 1953.

32. Robbins, P. S.; Alm, S. R.; Armstrong, A. L.; Averill, A. L.; Baker, T. C.; et al. Trapping *Phyllophaga* Spp. (Coleoptera: Scarabaeidae: Melolonthinae) in the United States and Canada Using Sex Attractants. *Journal of Insect Science* **2009**, *6*, 1–124.

33. Buss, E. A. Flight Activity and Relative Abundance of Phytophagous Scarabs (Coleoptera: Scarabaeidae) in Florida. *Florida Entomologist* **2006**, *89*, 32–40.

34. Price, G. Y. *Relationship between Japanese Beetle (*Popillia japonica *Newman) Larval Density, and Microbial Community Structure in Soil Microcosms*. MS thesis; Purdue University, 2017.

35. Billeisen, T. L.; Brandenburg, R. L. Biology and Management of the Sugarcane Beetle (Coleoptera: Scarabaeidae) in Turfgrass. *Journal of Integrated Pest Management* **2014**, *5* (4). https://doi.org/10.1603/IPM14008.

36. Vittum, P. J. Annual Bluegrass Weevil. In *Handbook of Turfgrass Insects;* Brandenburg, R. L., Freeman, C. P., Eds., *2nd* ed.; Entomological Society of America: Maryland, 2012.

37. Dupuy, M. M.; Ramirez, R. A. Biology and Management of Billbugs (Coleoptera: Curculionidae) in Turfgrass. *Journal of Integrated Pest Management* **2016**, *7* (1). https://doi.org/10.1093/jipm/pmw004.

38. Huang, T. I.; Buss, E. A. Billbug (Coleoptera: Curculionidae) Species Composition, Abundance, Seasonal Activity, and Development Time in Florida. *Journal of Economic Entomology* **2009**, *102,* 309−314.

39. Doskocil, J. P.; Brandenburg, R. L. Billbug (Coleoptera: Curculionidae) Life Cycle and Damaging Life Stage in North Carolina, with Notes on Other Billbug Species Abundance. *Journal of Economic Entomology* **2012**, *105,* 2045−2051.

40. Reynolds, D. S.; Reynolds, W. C.; Brandenburg, R. L. Overwintering, Oviposition, and Larval Survival of Hunting Billbugs (Coleoptera: Curculionidae) and Implications for Adult Damage in North Carolina Turfgrass. *Journal of Economic Entomology* **2016**, *109,* 240−248.

41. Duffy, A. G.; Powell, G. S.; Zaspel, J. M.; Richmond, D. S. Billbug (Coleoptera: Dryophthoridae: *Sphenophorus* spp.) Seasonal Biology and DNA-Based Life Stage Association in Indiana Turfgrass. *Journal of Economic Entomology* **2018**, *111,* 304−313.

42. Diaz, M. D. C.; Peck, D. C. Overwintering of Annual Bluegrass Weevils, *Listronotus maculicollis,* in the Golf Course Landscape. *Entomologia Experimentalis et Applicata* **2007**, *125,* 259−268.

43. Cowles, R.; Alm, S. R.; Villani, M. G. Selective Toxicity of Halofenozide to Exotic White Grubs (Coleoptera: Scarabaeidae). *Journal of Economic Entomology* **1999**, *92,* 427−434.

44. Bixby, A.; Alm, S. R.; Power, K.; Grewal, P. Susceptibility of Four Species of Turfgrass-Infesting Scarabs (Coleoptera: Scarabaeidae) to *Bacillus Thuringiensis* Serovar *Japonensis* Strain Buibui. *Journal of Economic Entomology* **2007**, *100,* 1604−1610.

45. Frank, J. H.; Parkman, J. P. Integrated Pest Management of Pest Mole Crickets with Emphasis on the Southeastern USA. *Integrated Pest Management Reviews* **1999**, *4,* 39−52.

46. Hudson, W. G.; Braman, S. K.; Abraham, C. M. Mole Crickets. In *Handbook of Turfgrass Insects;* Brandenburg, R. L., Freeman, C. P., Eds., *2nd* ed.; Entomological Society of America: Maryland, 2012.

47. Held, D. W.; Cobb, P. G. *Biology and Control of Mole Crickets, ANR-0176;* Alabama Cooperative Extension System: Auburn, AL, 2016; p 9.

48. Xu, Y.; Held, D. W.; Hu, X. P. Potential Negative Effects of Earthworm Prey on Damage to Turfgrass by Omnivorous Mole Crickets (Orthoptera: Gryllotalpidae). *Environmental Entomology* **2012**, *41* (5), 1139−1144.

49. Xu, Y.; Held, D. W.; Hu, X. P. Dietary Choices and Their Implications for Survival and Development of Omnivorous Mole Crickets (Orthoptera: Gryllotalpidae). *Applied Soil Ecology* **2013**, *71,* 65−71.

50. Villani, M. G.; Allee, L. L.; Preston-Wilsey, L.; Consolie, N.; Xia, Y.; Brandenburg, R. L. Use of Radiography and Tunnel Castings for Observing Mole Cricket (Orthoptera: Gryllotalpidae) Behavior in Soil. *American Entomologist* **2002**, *48,* 42−50.

51. Bailey, D. L.; Held, D. W.; Kalra, A.; Twarakavi, N.; Arriaga, F. Biopores from Mole Crickets (*Scapteriscus* spp.) Increase Soil Hydraulic Conductivity and Infiltration Rates. *Applied Soil Ecology* **2015**, *94,* 7−14.

52. Brandenburg, R. L.; Yulu, X.; Schoeman, A. S. Tunnel Architectures of Three Species of Mole Crickets (Orthoptera:Gryllotalpidae). *Florida Entomologist* **2002**, *85*, 383–385.

53. Ulagaraj, S. M. Mole Crickets: Ecology, Behavior, and Dispersal Flights (Orthoptera: Gryllotalpidae: *Scapteriscus*). *Environmental Entomology* **1975**, *4*, 265–273.

54. Hayslip, N. C. Notes on Biological Studies of Mole Crickets at Plant City, Florida. *Florida Entomologist* **1943**, *26*, 33–46.

55. Rethwisch, M. D.; Baxendale, F. P.; Dollison, D. R. First Report of Northern Mole Cricket Damage on a Nebraska Golf Course (Orthoptera: Gryllotalpidae). *Journal of the Kansas Entomological Society* **2009**, *82*, 103–105.

56. Held, D. W.; Xu, Y. Field Performance and Consumption of Indoxacarb Bait for Control of Mole Crickets (*Scapteriscus* spp.) in Turfgrass. *Crop Forage & Turfgrass Management* **2015**, *1* (1). cftm2015.0132.

57. Foldi, I. Ground Pearls: a Generic Revision of the Margarodidae *Sensu Stricto* (Hemiptera: Sternorrhyncha:Coccoidea). *Annals Society Entomology France (n.s.)* **2005**, *41*, 81–125.

58. Spink, W. T.; Dogger, J. R. Chemical Control of the Ground Pearl, *Eumargarodes laing*. *Journal of Economic Entomology* **1961**, *54*, 423–424.

59. Hoffman, E.; Smith, R. L. Emergence and Dispersal of *Margarodes meridionalis* (Homoptera: Coccoidea) in Hybrid Bermudagrass. *Journal of Economic Entomology* **1991**, *84*, 1668–1671.

60. Walker, P. W.; Allsopp, P. G. Factors Influencing Populations of *Eumargarodes laingi* and *Promargarodes* spp. (Hemiptera: Margarodidae) in Australian Sugarcane. *Environmental Entomology* **1993**, *22*, 362–367.

61. Kerr, S. H. Ground Pearls in Florida. *Florida Turf Association Bulletin* **1957**, *4* (1), 3.

62. *Grounds Maintenance*, 2013. http://www.grounds-mag.com/mag/grounds_maintenance_dont_give_ground/.

63. Hertl, P. Ground Pearls. In *Handbook of Turfgrass Insects;* Brandenburg, R. L., Freeman, C. P., Eds., *2nd* ed.; Entomological Society of America: Maryland, 2012.

64. Hodgson, E. W.; Roe, A. H. *Cranberry Girdler, ENT-42-07;* Utah State University Extension, 2007; p 2.

65. Salisbury, S. E.; Anderson, N. P. Grass Seed Pests. In *Pacific Northwest Insect Management Handbook [online];* Hollingsworth, C. S., Ed.; Oregon State University: Corvallis, Oregon, 2019.

66. Stahnke, G.; Antonelli, A. L.; Peck, D. C. European and Common Crane Flies. In *Handbook of Turfgrass Insects;* Brandenburg, R. L., Freeman, C. P., Eds., *2nd* ed.; Entomological Society of America: Maryland, 2012.

67. Peck, D. C.; Olmstead, D. L.; Petersen, M. J. Pest Status of Invasive Crane Flies in New York Turfgrass and the Repercussions for Regional Plant Protection. *Journal of Integrated Pest Management* **2010**, *1*. E1–E8J.

68. Jackson, D. M.; Campbell, R. L. *Biology of the European Crane Fly* In: *Tipula paludosa* Meigen, in Western Washington (Tipulidae: Diptera)*;* Washington State University Technical Bulletin no. 81, 1975; p 23.

69. Blackshaw, R. P.; Coll, C. Economically Important Leatherjackets of Grasslands and Cereals: Biology, Impact, and Control. *Integrated Pest Management Reviews* **1999**, *4*, 143–160.

70. Dawson, L. A.; Grayston, S. J.; Murray, P. J.; Pratt, S. M. Root Feeding Behavior of *Tipula paludosa* (Meig.) (Diptera: Tipulidae) on *Lolium perenne* (L.) and *Trifolium repens* (L.). *Soil Biology and Biochemistry* **2002**, *34*, 609–615.

71. Peck, D. C.; Hoebeke, E. R.; Klaus, C. Detection and Establishment of the European Crane Flies, *Tipula paludosa* Meigen and *Tipula oleracea* L. (Diptera: Tipulidae) in New York: A Review of Their Distribution, Invasion History, Biology, and Recognition. *Proceedings of the Entomological Society of Washington* **2006**, *108*, 985–994.

72. Gelhaus, J. K. Larvae of the Crane Fly Genus *Tipula* in North America (Diptera: Tipulidae). *University of Kansas Science Bulletin* **1986**, *53* (3), 121–182.

73. Rao, S.; Liston, A.; Crampton, L.; Takeyasu, J. Identification of Larvae of Exotic *Tipula paludosa* (Diptera: Tipulidae) and *T. oleracea* in North America Using Mitochondrial *cytB* Sequence. *Annals of the Entomological Society of America* **2006**, *99*, 33–40.

74. López, R.; Potter, D. A. Biodiversity of Ants (Hymenoptera: Formicidae) in Golf Course and Lawn Turf Habitats in Kentucky. *Sociobiology* **2003**, *42*, 701–714.

75. Kunkel, B. A.; Held, D. W.; Potter, D. A. Impact of Halofenozide, Imidacloprid, and Bendiocarb on Beneficial Invertebrates and Predatory Activity in Turfgrass. *Journal of Economic Entomology* **1999**, *92* (4), 922–930.

76. Barden, S. A.; Held, D. W.; Graham, L. C. Lack of Interactions between Fire Ant Control Products and White Grubs (Coleoptera: Scarabaeidae) in Turfgrass. *Journal of Economic Entomology* **2011**, *104* (6), 2009–2016.

77. López, R.; Held, D. W.; Potter, D. A. Management of a Mound-Building Ant, *Lasius neoniger* Emery, on Golf Course Putting Greens Using Delayed Action Baits or Fipronil. *Crop Science* **2000**, *40*, 511–517.

78. Vogt, J. T.; Wallet, B.; Coy, S. Dynamic Thermal Structure of Imported Fire Ant Mounds. *Journal of Insect Science* **2008**, *8*, 31.

79. Rhoades, R. B.; Stafford, C. T.; James, F. K., Jr. Survey of Fatal Anaphylatic Reactions to Imported Fire Ant Stings. *The Journal of Allergy and Clinical Immunology* **1989**, *84*, 159–162.

80. Lard, C. F.; Hall, C.; Salin, V.; Vinson, B.; Cleere, K. H.; Purswell, S. *The Economic Impact of the Red Imported FIre Ant on the Homescape, Landscape, and the Urbanscape of Selected Metroplexes of Texas* In: *Fire Ant Economic Research Report # 99-08*; Texas A&M University, 1999; p 64.

81. Potter, D. A. Natural Enemies Reduce Pest Populations in Turf. *USGA Green Section Record* **1992**, *30* (6), 6–10.

82. Zenger, J. T.; Gibb, T. J. Identification and Impact of Egg Predators of *Cyclocephala lurida* and *Popillia japonica* (Coleoptera: Scarabaeidae) in Turfgrass. *Environmental Entomology* **2001**, *30*, 425–430.

83. López, R.; Potter, D. A. Ant Predation on Eggs and Larvae of the Black Cutworm and Japanese Beetle in Turfgrass. *Environmental Entomology* **2000**, *29*, 116–125.

84. Braman, S. K.; Pendley, A. F. Relative and Seasonal Abundance of Beneficial Arthropods in Centipede Grass as Influenced by Management Practices. *Journal of Economic Entomology* **1993**, *86*, 494–504.

85. Frank, S. D.; Shrewsbury, P. M. Effect of Conservation Strips on the Abundance and Distribution of Natural Enemies and Predation of *Agrotis ipsilon* (Lepidoptera: Noctuidae) on Golf Course Fairways. *Environmental Entomology* **2004**, *33*, 1662–1672.

86. Hong, S. C.; Held, D. W.; Williamson, R. C. Generalist Predators and Predation of Black Cutworms, *Agrotis ipsilon* in Close Mown Creeping Bentgrass. *Florida Entomologist* **2011**, *94* (3), 714-175.

87. Joseph, S. V.; Braman, S. K. Influence of Plant Parameters on Occurrence and Abundance of Arthropods in Residential Turfgrass. *Journal of Economic Entomology* **2009**, *102*, 1116–1122.

88. Carstens, J.; Heng-Moss, T.; Baxendale, F.; Gaussoin, R.; Frank, K.; Young, L. Influence of Buffalograss Management Practices on Western Chinch Bug and its Beneficial Arthropods. *Journal of Economic Entomology* **2007,** *100,* 136–147.

89. Held, D. W.; Wheeler, C.; Abraham, C.; Pickett, K. Paper Wasps (*Polistes* spp.) Attacking Fall Armyworm Larvae (*Spodoptera: Frugiperda*) in Turfgrass. *Applied Turfgrass Science* **2008;** https://doi.org/10.1094/ATS-2008-0806-01-RS.

90. Obeysekara, P. T.; Legrand, A.; Lavigne, G. Use of Herbivore-Induced Plant Volatiles as Search Cues by *Tiphia vernalis* and *Tiphia popilliavora* to Locate Their Below-Ground Scarabaeid Hosts. *Entomologia Experimentalis et Applicata* **2014,** *150,* 74–85.

91. Rogers, M. E.; Potter, D. A. Kairomones from Scarabaeid Grubs and Their Frass as Cues in Below-Ground Host Location by the Parasitoids *Tiphia vernalis* and *Tiphia pygidialis*. *Entomologia Experimentalis et Applicata* **2002,** *102,* 307–314.

92. Bixby-Brosi, A. J.; Potter, D. A. Endophyte-mediated Tritrophic Interactions between a Grass-Feeding Caterpillar and Two Parasitoid Species with Different Life Histories. *Arthropod-Plant Interactions* **2012,** *6,* 27–34.

93. Callcott, A.-M. A.; Porter, S. D.; Weeks, R. D., Jr.; Graham, L. C.; Johnson, S.; Gilbert, L. E. Fire Ant Decapitating Fly Cooperative Release Programs (1994–2008): Two *Pseudacteon* Species, *P. tricuspis* and *P. curvatus*, Rapidly Expand across Imported Fire Ant Populations in the Southeastern United States. *Journal of Insect Science* **2010,** *11,* 19.

94. Rogers, M. E.; Potter, D. A. Potential for Sugar Sprays and Flowering Plants to Increase Parasitism of White Grubs by Tiphiid Wasps (Hymenoptera: Tiphiidae). *Environmental Entomology* **2004,** *33,* 619–626.

95. Joseph, S. V.; Braman, S. K. Predatory potential of Geocoris spp. and Orius insidiosison fall armyworm in resistant and susceptible turf. *Journal of Economic Entomology* **2009,** *102,* 1151–1156.

96. Redmond, C. T.; Williams, D. W.; Potter, D. A. Comparison of scarab grub populations and associated pathogens and parasitoids in warm- or cool-season grasses used on transitional zone golf courses. *Journal of Economic Entomology* **2012,** *105* (4), 1320–1328.

97. Williamson, R. C.; Obear, G. R. Organic Fertilizer Deters Vertebrate Pests from White Grub-Infested Turf. *International Turfgrass Society Research Journal* **2017,** *13,* 1–3.

Chapter 7

Insects and mites attacking woody and herbaceous plants

Niches for insect and mites

The species diversity and growth habits of trees, shrubs, and herbaceous annual and perennial plants in urban landscapes provide unique opportunities (niches) for insects and mites to colonize these parts (Fig. 7.1). Plant leaves or needles are the main site of photosynthesis for plants. They are also one of the more nutritious plant parts for insects. Leaves or needles host insects or plant-feeding mites that feed from the surface by chewing the foliage or sucking plant sap. There are leaf or needleminers that develop as larvae inside those tissues. Miners and borers represent insects known as endophages. Endophagy is the habit of an insect life stage (usually immatures) to feed and develop inside rather than outside of the plant. Miners can feed between the upper and lower surface of leaves or needles, or they can feed between the layers of outer bark (cambium miners) in the stems of woody plants. Miners are protected from external forces and may manipulate the plant to nutritionally enhance mined leaves (1). Leafmining is mainly esthetic but may reduce photosynthesis and growth (2). Populations of a few leafminers (birch leafminer) may be so great that most leaves are infested, resulting in defoliation. Borers are insects with chewing mouthparts that include bud borers (budworms), petiole borers, shoot borers, root borers, trunk or stem borers, twig girdlers, and pruners. Borers and miners create a gallery or mine, a cavity inside the plant, where the larva(e) develops. Most endophagous insects are in the orders Coleoptera, Lepidoptera, Diptera, and Hymenoptera. Borers that are beetles and weevils (Coleoptera) or larvae of moths (Lepidoptera) are the most common in urban landscapes. Some borers, such as emerald ash borer, have been introduced to the United States and have caused regional extinction of some tree species or genera (3). Borer infestations are associated with progressive plant decline, but a few species, mostly bark and ambrosia beetles, are able to weaken or kill trees in a large coordinated attack or in one year. Borer larvae are the damaging life stage for most species. There are some insect species that are only endophagous for

Urban Landscape Entomology. https://doi.org/10.1016/B978-0-12-813071-1.00007-5

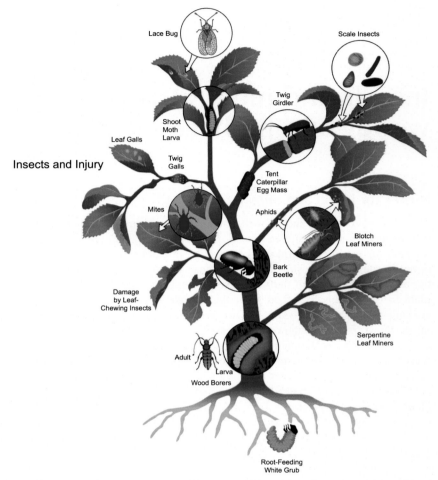

FIGURE 7.1   Ornamental plants provide unique opportunities (niches) for insects and mites to colonize above ground plant parts. *Adapted from Johnson, W. T.; Lyon, H. H. Insects that feed on trees and shrubs; 2ⁿᵈ Ed; Cornell University Press: Ithaca, 1991.*

part of one life stage. Spruce budworm (a tortricid moth), for example, consumes spruce foliage and then enters the vegetative buds. Once new needles emerge, the larvae then feed externally on them. Birch casebearers and juniper webworms feed as leafminers as young caterpillars, then feed outside on foliage as older larvae (4). Borers are also confusing because the common names rarely provide insight into identification. For example, dogwood borer can refer to either a beetle (*Oberea*) that infests twigs or a moth that infests the trunk (4). Furthermore, species with the same habits such as leafmining can occur on the same plant. The herbaceous plant columbine has two related leafminers that infest native and nonnative plants (5).

Plant galls are nonreversible, abnormal plant growths and represent a unique niche unlike the others mentioned so far. Plant galls can be caused by nematodes, pathogens, mites, and insects with insects being the more common causes or inducers of plant galls (4, 6, 7). An estimated 15,000 insects can induce galls on plants, however, interactions only a few species of gall inducers are well understood (7). Once induced, the plant is stimulated to overgrowth or enlargement of plant tissues, which results in a leaf (Fig. 7.2) or stem (Fig. 7.3) gall (Box 7.1). Galls do not return to normal tissue once induced by insects. Gall inducers hijack the plants' genetic machinery to force the plant to create a unique resource for feeding (1). These new resources somewhat resemble plant structures but are unique plant organs only made possible through an interaction with a gall inducer. The greatest number and diversity of plant galls, about 98% of known plant galls, occur on the leaves of deciduous plants. Specifically, in North America, about 50% of galls are

FIGURE 7.2 About 98% of known plant galls occur on the leaves of deciduous plants. These are examples of leaf galls.

FIGURE 7.3 Most stem galls on woody and herbaceous plants are merely aesthetically damaging. However, damage from leaf and stem galls by the Erythrina gall wasp can severely deform leaves and stems. When infestations are heavy, the trees may die.

reported on plants in the Fagaceae (oaks, beech, hickory), and 15% on rosaceous plants (*6*). Because these plant families are well represented in urban landscapes (*8—10*), the probability of encountering plant galls on ornamental plants is high. The insect gall inducer is considered an effector and not just an infector. Effectors transmit molecules, unknown in most cases, to the plant like a virus, then the plant has the virus independent of continued feeding by the insects. Instead, gall inducers transmit molecules called effectors to plants through feeding or oviposition. If the gall inducer dies, the plant does not continue to make the gall (*1,7*). Galls range in complexity from the chambered, woody stem galls of the horned oak gall to the simpler leaf rolls of Cuban laurel thrips. Galls, especially stem galls, can alter plant architecture. Large stem galls on oak made by wasps or club galls on dogwood can kill the twig or branch from the gall to the terminal (*11,12*). Because of the nature of the host interaction, all galls are host specific. The uniqueness and oddity of woody and herbaceous plant galls make information on galls and the gall inducers relatively easy to find. Books (*4,13—15*) and field guides often list information on galls with the host plant. Unfortunately, that is about the extent of what is known ecologically or morphologically about galls. Only in certain cases do we know how galls interact with the plant, natural enemies, and other herbivores (*16*).

> **BOX 7.1 Potential reasons for gall induction**
>
> Research with gall inducer suggests a few hypotheses to explain why insects and mites would have such intricate feeding behaviors (*1,6,7,17*):
>
> 1) Nutrition: Galled tissues have better nutritional value for insects than non-galled tissue. This may also explain why other plant feeding insects and mites will feed and develop inside galls (*4,18*). These freeloading arthropods are called inquilines. They can inadvertently kill some of the gall inducers.
> 2) Microenvironment: The abiotic conditions inside the gall are less variable than outside.
> 3) Protection: Losses due to natural enemies should be less inside a gall structure than outside.
> 4) Mimicry: Many leaf and stem galls resemble other insects or even fruit like acorns on oak trees. Appearing as other plant structures or other insects may reduce losses to inquilines.

Many of the insects and mites associated with ornamental plants in urban landscapes have common names, which include a plant name. As previously discussed, insect herbivores are mostly specialists so it is reasonable to think a plant name may indicate host plant specialization. Unfortunately, the common names do not consistently provide information about whether the insect is a specialist or a generalist. For example, the moth called dogwood borer is able to develop in many different plants including stem gall tissues (*18*). Green peach aphids can develop on hundreds of plants during the summer, whereas crapemyrtle aphids only develop on crapemyrtle. In these examples, the plant name is not communicating whether or not the insect is a host specialist or a generalist. Instead, it is likely indicating the host plant where the aphids will overwinter as eggs.

A majority of species in the insect orders Lepidoptera, Phasmida, Thysanoptera, Hemiptera, and Orthoptera are plant feeders. Plant-feeding beetles, larvae of flies, and sawflies round out the list of orders that can cause plant damage (Table 7.1). The species diversity and complex behaviors make the study and diagnosis of ornamental pests and pest management more difficult than with pests of turfgrass in the landscape. A number of published books (*4,13,15,19*) use diagnostic images and descriptions of signs, symptoms, and damage to aid identification of ornamental pests. Books and websites that catalog the extensive diversity are great resources, but it is not possible in one chapter or even some books to provide an answer to every, "What is this?" question. For these reasons, the rest of this chapter organizes the insect and mite diversity according to the niches (borers, gall inducers) used by each group. Each section will include citations for identification resources, and when appropriate, diagnostic characters for field identification of certain groups are provided.

TABLE 7.1 Feeding niches for eight insect orders and two families of mites that use ornamental plants (4).

Order\family	Feeding location or niche	Examples of plant feeders
Lepidoptera	Defoliators*, twig, trunk and shoot borers, leafminers, cambium borer\bark miner	Bagworms, dogwood borer, maple shoot borer, azalea leafminer, *Marmara* barkminer
Hymenoptera	Defoliators*, stem and shoot borers, leafminers, budminers, gall inducers*, parasitoids*, predators	Sawflies, roseslugs, leafcutter bees and ants, *Hartigia* cane borer, birch leafminer, cynipid wasps (gall inducers), *Pleroneura* budminer
Coleoptera	Defoliators*, stem, shoot, and root borers, leaf miners, gall inducers, budminers, predators*	Japanese beetles, flea beetles, palm weevils, southern and mountain pine beetles, gall inducers are uncommon
Diptera	Leafminers*, gall inducers*, cambium borer\bark miner, parasitoids, predators	Holly leafminer, dogwood club gall midge, *Phytobia* barkminers
Orthoptera	Defoliators	Differential grasshopper, lubber grasshopper, tree crickets, katydids
Phasmida	Defoliators	Twostriped walkingstick
Hemiptera	Sap suckers, gall inducers, predators	Cicadas, aphids, scale insects, leafhoppers, adelgids, lace bugs
Thysanoptera	Sap suckers, gall inducers, predators	Western flower thrips, Cuban laurel thrips (galls)
Tetranychidae	Sap suckers	Spider mites
Eriophyidae	Sap suckers, gall inducers	*Phyllocoptes fructiphilus* (mite involved with rose rosette), elm finger gall, maple spindle gall

* More common feeding locations or niches for these orders and families.

Sap-sucking insects and mites

The sap-sucking insects occur in the orders Hemiptera and Thysanoptera (thrips), and the plant-feeding mites occur in four families of which two (Tetranychidae and Eriophyidae) will be discussed here. The mouthparts of these insects and mites are modified into a beak to enable liquid feeding. The plant-parasitic hemipterans (Sternorrhyncha) include insects that are not typically found independent from their host plants. These include aphids, whiteflies, and scale insects. Scale insects are also represented by many families, but this chapter will only include the four most common in urban landscapes (armored scales, soft scales, giant scales, and felt\bark scales). Plant parasitic hemipterans, mites, and thrips are the smallest sized arthropods in the urban landscape. Their mouthparts and sometimes their body features, in general, are not easily seen without a hand lens or microscope. Spider mites (arachnids) can co-occur on plants with small insects (Fig. 7.4), and situations may require separation of spider mites from other small insects. Spider mites will have four pairs of legs, a ticklike body shape, and will never have wings. Furthermore, spider mites can use silk to move around the plant. The presence of silk coupled with the body form directs a diagnosis to spider mites. As we learned in Chapter 6, eriophyid mites are microscopic and are not observable in the field without a strong hand lens. The free-living, plant-feeding insects (Auchenorrhyncha) include leafhoppers and cicadas. These are larger than the plant parasitic hemipterans with free-living adults that are generically lumped under the common names of plant or leaf hoppers. The immature stages are found associated with their host plants and bear almost no resemblance to the adults. The true bugs (Heteroptera) include both predatory and plant-feeding sap-sucking insects. They are free living as adults and immatures, and both

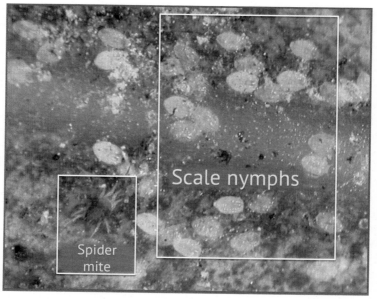

FIGURE 7.4 Small mites and insects can co-occur on plants. A plant-feeding spider is shown next to nymphs of false oleander scale.

life stages bear some resemblance to one another and co-occur on plants. The small size and cryptic habits of sap-sucking insects explain why so many are incidentally moved between states or internationally on shipments of plants. Historic introductions of armored scales such as San Jose scale (20) and the recently introduced spotted lantern fly (a planthopper) (21) are evidence of the destructive and economic impacts of sap-sucking insects once they enter a region devoid of their endemic natural enemies.

Sap-sucking insects and mites are either free living or gall formers (Table 7.1), but never borers or miners. Feeding by adults or immatures causes the induction of galls, most of which are leaf galls. Sap-sucking insects and mites vary in their use of different plant sap (phloem, xylem, and cell contents) and how they access their liquid diet. First, sucking mouthparts are not all equal. Within the beak of all sap-sucking insects or mites are thread-like structures called stylets. Sap-sucking insects have four, retractable stylets that come together to form a food canal and salivary canal (22). Retractable here means they can be inserted and withdrawn. Relative to the other hemipterans, the auchenorrhynchans (hoppers and cicadas) have a much shorter beak than the other hemipterans. Thrips (Thysanoptera) similarly have a short structure called a mouthcone and asymmetrical mouthparts: a single left mandibular stylet and two well-developed maxillary stylets (three total) (23). They puncture the epidermis with the mandibular stylet and then probe the epidermal and mesophyll cells of leaves and buds. Like hemipterans, mouthparts also contain food and salivary canals. Feeding causes characteristic silvery or necrotic damage on leaves, flowers, or buds. Spider mite mouthparts (two chelicerae) have a single, retractable stylet and one canal. Eriophyid mites have similar mouthparts to spider mites, but the stylets are not retractable. Insect stylets are about five to eight times longer than the stylets of twospotted spider mite adults and nymphs (24). The length of the stylets influences the type of plant sap that can be accessed. Eriophyid mites have short stylets (7—30 microns long) less than half the diameter of a human hair and therefore can only access the contents of plant epidermal cells (25). Twospotted spider mites have longer stylets (100—150 microns long), about 1.5—2 times the diameter of a human hair, so they can access the parenchyma cells just below the epidermal cells. Mites generally remove cell contents, 1 cell at time, but can damage about 20 adjacent cells in about a minute (24). The cells fed upon will collapse and eventually develop the symptom known as stippling. Spider mite feeding does not immediately produce stippling, but that symptom appears a few days after feeding (24,26).

With longer stylets, insects access cell contents, xylem, or phloem. If honeydew or sooty mold is present, the sap-sucking insects are using phloem. Phloem sap is accessed by aphids, mealybugs, whiteflies, giant scales, and soft scales. Psyllids (pronounced sill-ids), adelgids, treehoppers, and leafhoppers have different species in the family that use either xylem or phloem (27) or some species may access both xylem and phloem (28). Armored scales access either xylem or consume contents of parenchyma cells. Xylem use appears more common for armored scales on plant stems and armored scales on leaves feed on cell contents (*29–31*). The heteropterans (lace bugs, stink bugs, seed bugs, plant bugs) consume cell contents through a lacerate and flush habit. Lace bugs, for example, feed from the undersides of the leaves, with their mouthparts (stylets) entering through the plant stomates. The stylets then weave through the leaf piercing the parenchyma cells below the upper epidermis (*32*). They rupture cell contents by repeatedly inserting and retracting the stylets. As the stylets move about, they also pump a watery saliva into the plant. The mixture of water and dissolved cells is then consumed (*33,34*). When the cell chlorophyll content is reduced as a result of sap feeding, the infested plant will have a decrease in photosynthetic rate (*29,32,35*). Rupturing cells can produce chlorotic or necrotic (dead) areas around the feeding site. In certain armored scales on leaves, this appears like a yellow halo or patch beneath and adjacent to the scales (Fig. 7.5). For certain plant bugs, these cells die and superficially resemble feeding or skeletonization caused by insects with chewing mouthparts. In addition to stippling, necrosis, or chlorosis, cupping or curling of infested leaves may also result. The loss of chlorophyll from cell content feeders causes a decrease in photosynthetic rates as much as 50%–60% on infested plants (*32,35*). This section on feeding is not entirely academic; the type of plant sap used can influence exposure to systemic insecticides (discussed in Chapter 9) or pest life history. Female soft scales and giant scales, phloem feeders, generally have hundreds to thousands of eggs per female compared with 10–30 per female in some armored scales (*4*). Similarly, xylem feeding among hemipterans (leafhoppers, tree-hoppers, and spittlebugs) is the only known way that the bacterium, *Xylella*

FIGURE 7.5 Chlorotic areas or halos are common in plant leaves infested with armored scales.

fastidiosa, is spread from plant to plant. These bacteria are the causal agents of bacterial leaf scorch in landscape trees and Pierce's disease, a fatal disease of grapes (*36*).

Wings are typically the outward feature that an insect is an adult form. Aphids, adelgids, phylloxera, mealybugs, and scale insects, however, have winged and nonwinged forms of the same species as adults (Fig. 7.6). In some aphids, such as crapemyrtle aphids, all adults are winged (*37*), but winged forms of many aphids are only occasionally present. In scale insects and mealybugs, only the adult males, if present, have wings. Despite developing in close proximity, males need wings to find and mate with females. However, mating is not required for many of these sap suckers to produce viable offspring. Females have the ability to produce one sex from nonfertilized eggs (commonly but not always females) and the other sex if mated. The offspring from nonfertilized eggs are essentially clones. In urban landscapes, adult psyllids, thrips, leafhoppers, planthoppers, spittlebugs, cicadas, lace bugs, stink bugs, and plant bugs will have wings. Mites never have wings, so adults are only distinguished from immatures by size. Adult stages of sap-sucking insects and mites can vary behaviorally from the immature forms. Female scale insects, lace bugs, and twospotted spider mites generally feed more and cause more damage than immatures or males (*26,32,35*). Some developing nymphs like of Asian citrus psyllids may have greater feeding rates than adults (*38*). Nymphs of certain hoppers (treehoppers, leafhoppers, and spittlebugs) also have different host plants entirely for immature and adult forms. Different nutritional requirements of the adult and immature stages are one explanation for these observations (*39*). Male and female scale insects can also segregate on different plant parts as observed with armored scales. Female Euonymus scales are more common on stems, and males are more abundant on foliage

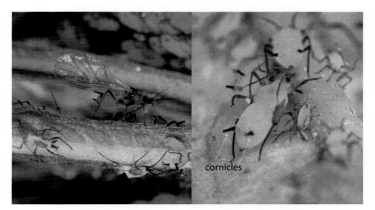

FIGURE 7.6 Winged and nonwinged adult forms are common in Sternorrhyncha. The cornicles shown on the adult and immature aphids are used to identify aphids to family (Aphididae).

(40). As scales do not disperse far, populations may adapt to local plant communities and form what are called demes (41). However, subsequent experiments with armored scales have failed to further support this as an explanation for local outbreaks in plant monocultures (41,42).

All sap-sucking insects and mites begin life as eggs deposited on, or embedded into the plant surface, inside a leaf fold (gall-inducing thrips), or that mature inside the female. Many of these species go through winter in temperate regions as eggs or adults. Nymphs begin life after hatching from eggs or after live birth (e.g., aphids). For scale insects and mealybugs, the mobile first instar stage is called a crawler. The crawler is the primary stage for dispersing between hosts or to different parts of the same plant. They can move between plants on wind or animals (43) or settle nearby where they emerge (41,44). When a favorable location is found, nymphs will begin feeding and are called settled crawlers. Settled crawlers also begin to secrete their waxy covering. In armored scales, the old exoskeleton is incorporated into the covered developed in the second and third instars (Fig. 7.7). Male scale insects while developing will complete extra molts yet typically have a covering that is smaller and a different shape compared with the females. It is not

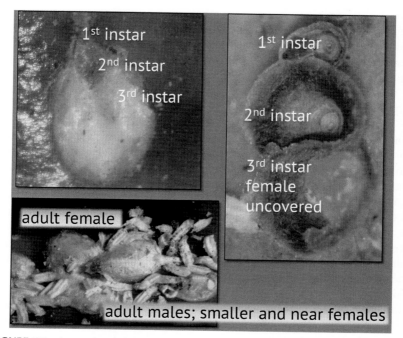

FIGURE 7.7 Armored scale insects appear differently as males and females. Because females expand the wax covering with each molt, the covering provides an indication of the instar. Note that the wax covering of armored scales can be lifted and separated from the insect.

uncommon for crawlers of armored scales to settle under or near the female, leading to layering or aggregation of scales. Mealybugs, giant scales, and some soft scales will retain legs after their first molt and can still move in the second instar stage. One or more nymph stages of mealybugs, psyllids, and whiteflies are flattened and appear scale-like.

For many sap-sucking insects on ornamental plants, it is entirely possible for all life stages, generation after generation, to develop on the same hosts for one or more years. However, there are species with slight variation on the typical life histories. Calico scale, *Eulecanium cerasorum* crawlers move to leaves on deciduous trees and shrubs, feed, and then molt into second instars. During fall, the second instars, which retain their legs, then move back to the branches to overwinter. In spring, the females will feed through the bark and produce copious amounts of honeydew (*45*). Some species of aphids and adelgids use alternating hosts. These species will usually have one species or genus of woody plant (primary host) on which they will pass the winter as eggs. Eggs are the overwintering life stage for aphids in nontropical areas; however, in most species, eggs are only used for over-wintering. The rest of the year they reproduce asexually by parthenogenesis. In the spring, females (stem mothers) hatch from the eggs and begin feeding. Later generations may remain on that host or disperse to other hosts (summer or secondary hosts). For aphids such as green peach aphids, this may include about 300 or more plant species. However, those aphids must find mates and locate a primary host by the end of the year, or they can locally go extinct. Adelgids in North America (*Pineus* or *Adelges* spp.) similarly alternate between spruce, on which they form galls, and other hosts on which they feed but do not induce galls. *Pineus* spp. use pine but still only form galls on spruce (*4*).

Thrips, treehoppers, and cicada cause damage from oviposition. These groups have sawlike ovipositors to cut a cavity into leaves or branches into which they deposit their eggs (*4*). Thrips oviposition can cause scars in young leaves or flowers only noticeable after they fully expand or open. Treehopper wounds can create opportunities for pathogens to enter woody plants. Female cicadas can create about 20 wounds per female during her lifetime (*4*). A weak spot in the branch may break resulting in a symptom called flagging, because the terminal end of the branch drops and hangs like a flag. Cicadas are either annual or periodical cicadas. Annual cicadas species have an adult emergence every year in the summer. Periodical cicada species only have a spring adult emergence every 13 or 17 years. The number of periodical cicadas that emerge is significantly more than for annual cicadas. For this reason, flagging damage is far worse in year of periodical cicada broods. Lace bugs (Tingidae) are leaf-feeding plant specialists that overwinter as either eggs or adults depending on the host plant. Lace bugs that use evergreen hosts (e.g., *Stephanitis* spp., azalea lace bug) overwinter as eggs deposited by females on leaves in the fall. In subtropical areas of the United States, all life stages can be found year-round.

Lace bugs that use deciduous hosts (e.g., *Coythuca* spp., oak lace bug) over-winter as adults (*4*). Eggs are deposited the following year after new foliage emerges. Lantana lace bug and some species that use herbaceous perennials likely overwinter as adults, but the seasonal biologies of these species of lace bugs are not as thoroughly studied.

Identification of the sap-sucking insects to order and suborder is possible with an image or a quality hand lens. Because of their size, mite identification is difficult, even for trained entomologists. Eriophyid mites are host specific, which can facilitate identification if the plant is correctly identified. One of the most serious eriophyid mites on ornamental plants is *Phyllocoptes fructiphilus*, the mite that vectors the virus that causes rose rosette disease in ornamental roses (*46*). These mites have been in North America for some time, but the extensive planting of landscape roses has provided opportunities for the mite and the virus to outbreak. Fortunately, most pests in urban landscapes are encountered with their host plant, and most horticulturalists, Master Gardeners, and landscape and turfgrass managers have training in plant identification. Family-level identification of insects is challenging but easier if you know the host plant. Most identification resources (called keys) use an image, or images accompanied by two different descriptive sentences called couplets. A couplet gives you an A or B choice. For those trained in plant identification, it may be easier to check with resources that list certain insect groups by their associated host plants. Online resources for thrips (*47*), aphids (*48*), scale insects (*49—51*), and mealybugs (*52*) are searchable using keywords such as host plant name. Here is an example using small insects discovered feeding on daylily foliage in the spring. You may recognize them as aphids by their cornicles (Fig. 7.6), or you may see a beak or honeydew and conclude they are sap-sucking insects (Hemiptera). For most situations, identification of insects to family is usually enough to provide the basic information for management, especially if an insecticide is being used. However, for aphids and other small insects further identification may be needed. The next step would be to determine the number of possible species of aphids (*48*) that feed on daylilies in spring. By checking for *Hemerocallis*, the Latin genus name for daylily, you get four results. Using the same resource, the geographic location of species can be determined. Of the four possible aphid species, only three are possible in North America, and only two are confirmed in North America. The aphids may be *Myzus hemerocallis,* but it is best to have the identification of species of aphids and all small insects and mites confirmed by an insect identifier. Aphids, adelgids, and phylloxera are similar in appearance to aphids, but adelgids and phylloxerans lack cornicles. Furthermore, adelgids only occur on conifers (pine, spruce, larch, fir), and phylloxerans only occur on deciduous trees. In urban landscapes, leaf galls on hickory (*Carya* spp.) are induced by phylloxerans. Aphids can occur on deciduous and evergreen hosts, and some species also induce galls.

Scale insects, psyllid, and whiteflies are among the most difficult to identify and manage insects in urban landscapes. The location and host of scale insects can assist with diagnostics. Certain species such as felt \ bark scales are only present on the trunk and branches. Yet, black pineleaf scale, hemlock scale, elongated hemlock scale, juniper scale, and pine needle scale only occur on leaves or needles. Armored scale insects have a wax covering that separates from the female (Fig. 7.7). Other scale insects are not separate from their waxy covers, and it is not possible to remove the covering without squashing the insect. Giant scales, mealybugs, felt scales, and some soft scales may have ovisacs, which are absent on armored scales. The wax covering does not completely cover the body in felt scales, and they do not retain legs as adults. Giant scales and mealybugs can be confused with one another because both are larger and flocculent with adults that retain legs, and females with ovisacs. The ovisac on giant scales is striated unlike the others. Soft scales in the genus *Pulivinaria* (cottony scales) also have an ovisac, but they are not striated and resemble the end of a cotton swab. Because most psyllids are specialists, the host plant will often be important to identification. Boxwood psyllids and lerp psyllids are two common psyllids in urban landscape that do not induce galls. In North America, the subtropical and tropical states have the most trouble with whiteflies because they overwinter outdoors or have continuous generations (*4*). The last immature life stage called a pupa is needed to confirm the identity of whiteflies because the adults are too generic for species identification. The immature stages of psyllids and whiteflies are flattened and resemble scale insects.

Defoliators and insects with chewing mouthparts

The remaining groups of insects have hardened mouthparts called mandibles to chew or skeletonize leaves. Mandibles are not exclusive to plant-feeding insect (also in many predatory insects), but they are required to shred and shear the leaf blades. Chewing mouthparts alone are not helpful for identification because the majority of insect species have mandibles in one or more life stages. Therefore, chewed leaves initially just tell you the insect is in one of six insect orders (Table 7.1). Defoliation is the major niche for these insects with the exception of flies (Diptera). Flies can feed in fruits (e.g., spotted winged Drosophila), as leaf miners or in roots, but are not defoliators of ornamental plants. Walking sticks (Phasmida), tree crickets, and grasshoppers (Orthoptera) are leaf consumers of ornamental plants in urban landscapes. The four orders with immatures that are larvae (Coleoptera, Lepidoptera, Diptera, and Hymenoptera) also share gall inducing, leafmining/needlemining, and stem and trunk boring habits. Hymenoptera are most commonly encountered in urban landscapes as defoliators (sawflies), gall inducers (gall wasps or sawflies), and a few important leafminers (birch leafminer). Leaf or complex stem galls encountered on oaks are likely caused by gall wasps (Cynipidae).

Leaf and stem galls on willows are induced by sawflies. Sawflies occur widely on various deciduous trees (birch, elm) in urban landscapes as leafminers, but not as needleminers on conifers (*4,53*). On roses three species of roseslug caterpillars, which are sawflies and not true slugs, can occur as defoliators. Cynipid wasps can induce galls on multiple plant parts, and stem borers in rose canes may also be sawfly larvae.

Dipteran (flies) larvae are commonly encountered as leafminers and gall inducers, with fewer species as cambium miners or twig borers. Feeding damage is caused by the larvae in all cases, but damage by females laying eggs into plant leaves is also common especially in leafminers. The leafminer flies (Agromyzidae) are mostly specialist leafminers (e.g., daylily leafminer, boxwood leafminer) with a few generalist leafminers. Leafminer flies are usually active in spring when young, tender growth is expanding on herbaceous and woody plants. The flies are small and not often noticed. The female lays eggs into the soft tissue, the eggs hatch, and the larvae make leafmines as they feed and develop. The leafminer flies that use woody hosts typically have one generation per year compared with multiple generations from those using herbaceous hosts (*4*). Native leafminer flies developing on their native host plants have many species of larval parasitoids (*5*). However, those parasitoids do not prevent mining because they do not act until larvae have hatched and are feeding. These leafminers usually overwinter in mines. Another family of flies (Cecidomyiidae) contains the gall inducers, the resin and needle midges in evergreens, and the few twig boring larvae (*4*). The gall midges are also spring active laying eggs into actively growing shoots or newly emerging leaves. Larvae feed inside the galls. At the end of the year, the larvae chew out of the galls and drop to the ground where they overwinter. The flies emerge from the pupae in the soil in spring to restart the life cycle (*4*). Because these midges also rely on newly expanding tissue, most of the gall midges also have one generation per year. Feeding by resin midges causes needles to droop and turn yellow. The midge is usually found in the glob of resin that accumulates on the infested tip. The midge overwinters in the bark and finishes development in the spring. Landscape pine trees surrounded by large stands of other pines are more likely to have damage (*4*). Cambium \ bark miners (Fig. 7.8) are fly larvae with a few species of moths. These insects are like leafminers except they create mines in the soft outer layers of bark. They are not common and are rarely ever damaging.

Lepidoptera are encountered in urban landscapes as defoliators, leaf rollers, tiers and folders, and trunk and branch borers, and leafminers. Outbreaks of defoliators of ornamental plants depend on the behavior of the females. Adult females are important because they select the host plant on which they lay eggs. Eggs are deposited on plants, usually leaves, as egg clusters or masses. Once the eggs hatch, the larvae begin feeding and develop through series of larval instars to pupation. Chapter 6 introduced the principle that larvae increase consumption of plant material dramatically as they age. Therefore, feeding by young larvae on ornamental plants is commonly missed.

FIGURE 7.8 Some caterpillars and immature flies develop as bark or cambium miners shown here on a young oak tree.

Azalea caterpillar eggs and larvae are present on host plants in the spring but often go unnoticed until later when older larvae feed more heavily. Once larvae begin feeding, they typically remain on that individual plant unless all leaves are consumed. Movement on or between plants and feeding by foliage feeding larvae make them more conspicuous to predatory insects and birds. For these reasons, feeding by larvae, especially when young, is commonly done underneath leaves or inside leaf rolls or folds, or inside the protection of silken bags or nests. The diagnostic use of silk was introduced in Chapter 5, but many species of eruciform larvae (for example, Monarch caterpillars, sphinx moth larvae) feed exposed on leaves without silken bags or tents. Just before they need to pupate, some species will wander away the host plant to complete their development. Wandering larvae can sometimes feed on other plants, as they travel to a site to pupate. This feeding is often minimal and rarely warrants control. Many of the snout moths that are leaf rollers (Pyraloidea) will overwinter in the dried stubble of their host plant. Some butterflies such as Monarchs and Swallowtails form an ornate pupal stage called a chrysalis sometimes attached to the host plant. All other pupal stages of eruciform larvae are bare and often drab colored in soil, or enclosed in a silken cocoon on or adjacent to the host plant. Eruciforms will overwinter as either pupae, larvae near pupation, or eggs on their host plants.

Larvae with leaf rolling, folding, and tier habits use silk to web together multiple leaves or needles or to roll a single leaf. About 200 species of caterpillars cause leaf folds of rolls (4). These are mostly in two families (Tortricidae and Pyralidae) of small moths. Once these leaf rolls and folds are opened, you will see the larva or pupa, or evidence of feeding like skeletonized leaves or frass. Leaf rolling and folding may be a behavioral adaptation to reduce plant toxins like coumarins or flavanoids, which would otherwise harm the developing larvae (54). These species typically overwinter as pupa within the leaf folds or rolls on the plant or in these plant materials that fall on the ground. The same two families of leaf rolling and folding caterpillars also have larva that infest the actively growing tips of trees (i.e., tip borers). Common examples include Nantucket pine tip moth and maple shoot borer. Like other borers, the caterpillar develops in the terminal end of the branch. This is problematic because when the terminal shoot is killed, the tree loses a central stem. As the tree grows, the lower branches grow more rapidly in the absence of a central stem causing the tree to grow wider than tall. Two main families (Sesiidae and Cossidae) of moths have caterpillars that are trunk and branch borers in woody plants (4,55−57). Forty species of clearwing moths (Sesiidae) and two carpenterworms (Cossidae) are considered pests of ornamentals. Females in both families must find trees and deposit egg(s). Females may use wounds on bark as locations for laying eggs (58) but do not appear attracted to common plant stress volatiles (*unpublished data*). These caterpillars hatch and chew inside the tree and develop. When near pupation, the larva chews an exit hole, if needed, and pupates at that hole. As the moth wriggles out at emergence, the pupal "skin" is left sticking out of the tree (Fig. 7.9). Wood-boring moths have one or two generations per year depending on species. Dogwood borer, *Synanthedon scitula*, and the lilac borer *Podosesia syringae*, are among the two most well-studied species in urban landscapes. The carpenterworm moth (*Prionoxystus robiniae*) and the wood leopard moth (*Zeuzera pyrina*) are the carpenterworm moths most associated with damage to a range of woody trees and shrubs. A species related to dogwood borer, *Synanthedon sequoiae*, are borers of pine in the western United States. They cause pitch (pine sap) to flow and develop as borer in those wounds creating an unsightly appearance but not typically being fatal (4).

FIGURE 7.9 When moths with wood-boring larvae emerge from woody plants, they leave behind pupal "skins" at the emergence site.

Beetles and weevils (Coleoptera) represent the same niches (defoliators, borers, leafminers, and a few gall inducers) already presented. Leaf-feeding beetles\weevils are from four of the largest families of beetles; Scarabaeidae (chafers, May/June beetles), Curculionidae (weevils), Chrysomelidae (leaf beetles and flea beetles), and Buprestidae (metallic wood—boring beetles). Insects in these families can be grouped into different list histories (4) based on where larvae feed (Box 7.2). Weevils, metallic wood—boring beetles, and longhorned beetles (Cerambycidae) are the main families of trunk, twig, and event root-boring species in ornamental plants. Weevils, and bark and ambrosia beetles are some of the more important borers in urban landscapes and especially responsive to trees under water stress or trees struck by lightning. The larvae of palm weevils (*Rhynchophorus* spp.) are the most destructive palm pests worldwide. Like the weevils in turfgrass, females create a notch for egg laying. Palm weevils notch the petiole near the base where the developing larvae can enter the palm trunk. Larvae develop in the crown or

---

BOX 7.2 Four common life histories of beetles that feed on leaves as adults

Life history (1) *Adult and larvae feed as defoliators of the same host plant.* This is the true for milkweed leaf beetles (*Labidomera*) and elm leaf beetles (*Xanthogaleruca*). It is typical for the larvae to leave the foliage and pupate either at the base of the plant or in the adjacent soil. These leaf-feeding beetles are specialists and will typically have two generations per year.

Life history (2) *Adults are leaf feeding, and larvae are leafminers.* With this life history, small (1.6 mm or 0.06 in.) adults overwinter near the host trees and begin feeding early in the spring. They deposit eggs in pockets along the midvein made by feeding. The larvae then feed as leafminers. The European elm flea weevil (*Orchestes alni*), the basswood leafminer (*Baliosus nervosus*), locust leafminer (*Odontota dorsalis*), and the yellow polar weevil are species with this life history. These species have either one or two generations per year (4,68). Despite the common names, few of these species are exclusive to those host plants.

Life history (3) *Adults are defoliators, and larvae are root feeders.* This is a common strategy for leaf-feeding beetles like strawberry rootworm, redheaded flea beetle, and all scarab beetles. Japanese beetles and May\June beetles are the adult forms of the white grubs mentioned in Chapter 6. May\June beetles are nocturnal, feeding on tree leaves. The damage symptoms are reported as appearing quickly, perhaps in one evening. May beetles typically feed until just the petiole and midvein of the leaf remain, although other insects may also have this feeding pattern.

Life history (4) *Adults are leaf or needle feeders, and larvae are stem borers.* This is a common strategy for metallic wood—boring beetles such as Emerald ash borer (3) and several genera of longhorned beetles (65). The extent of foliar damage from the adult beetles is usually minimal.

FIGURE 7.10 Bark and ambrosia beetles attack trees as adults. Ambrosia beetle attacks will result in an accumulation of sawdust at the base of the tree or sawdust toothpicks extruding from the tree. Larvae are small, nondescript grubs that develop inside a gallery.

trunk adjacent to foliage and kill susceptible palms trees (*59,60*). Unlike other wood-boring beetles that deposit eggs externally, female bark and ambrosia beetles enter the host tree to either lay eggs or mate and lay eggs. Bark beetles include southern and mountain pine beetles (*Dendroctonus* spp.), and engravers (*Ips* spp.) overwhelm the host tree using aggregation pheromones to recruit additional beetles to attack and for mating. Coincident with attack, females can introduce fungi to colonize and block the host vascular tissue. Larvae feed on the host tree and expand the initial galleries made by the female (*61*). Similarly, female ambrosia beetles common in urban landscapes (*Xylosandrus* spp.) colonize the host tree and produce a gallery. As they create the gallery, either sawdust will accumulate around the crown or sawdust "toothpicks" extrude from the entrance holes (Fig. 7.10). Females are typically mated before flight and lay eggs in the new gallery. Coincident with gallery production females introduce a fungus (*Ambrosiella* spp.) that grows in the gallery and provides nourishment for developing larvae. The ambrosia fungi are not pathogenic, but coincident infection with pathogenic fungi (*Fusarium*, *Phomopsis*, and *Nectria*) may occur in infested trees (*62,63*). Attacks by bark and ambrosia beetles can kill or severely weaken a tree. While five or more sticks per tree is suggested as a threshold for killing a tree 3 inches of less in caliper (*64*) attack rates and tree death have not been documented for a wide range of trees. Beetles in the family Cerambycidae use woody or herbaceous plants as well as roots (*Prionus* spp.)

to develop as larvae. Most use woody conifers and broadleaf trees (oaks) as larval hosts. Many species of longhorned beetles feed as adults on bark, pollen, nectar, or leaves. Certain genera and even subfamilies of longhorned beetles associate with or even require living hosts for larval development, and others associate with dead hosts (*65*). In North and Central America, the primary metallic wood—boring beetles (Buprestidae) in landscapes are species in *Chrysobothris* and *Agrilus*. *Agrilus* spp. includes the emerald ash borer and other native and invasive pest species (*3,57,66*). Herbaceous plants, woody shrubs, and trees are hosts for *Agrilus* spp. with oaks hosting the most species (*4,57*). *Agrilus* spp. will feed on, and occasionally do damage to, the foliage of trees that host their larvae. Woody plants, including gymnosperms, can be hosts for *Chrysobothris* and other genera of metallic wood—boring beetles. The metallic wood—boring beetles and longhorned beetles, along with a few moths like the juniper twig girdler, have the unique habit of twig girdling and pruning (*4*). The difference between girdlers and pruners is the life stage that cuts the branch of twig. Both borers develop as larvae inside smaller branches. Twig girdlers produce unique cuts around the circumference of the branches. Larvae overwinter and complete development inside the stem either on the tree or on the ground if the girdled twig breaks. Twig pruners are damaged by the larva leaving the end with a ragged cut, not the more precise cut of a twig girdler (*4*). All families of beetles listed above, except Scarabaeidae, include species of leafminers. The leafminers have adults that are leaf feeders on the same host as the larvae (life history 2). The three families of borers also include the uncommon species of gall-inducing beetles. Despite being documented worldwide, gall-inducing beetles are not well studied but form twig, branch, or root galls mainly on woody plants (*57,67*).

Stick insects (Phasmida), specifically the twostriped stick insects (Fig. 7.11), are rarely seen defoliators of a wide range of trees and shrubs. Most species do not fly, but they crawl and can move between hosts. The biology and number of generations are poorly defined. Eggs that look like seeds may be glued to plants or dropped by females to the ground as they feed. The eggs stage may take 1—2 years to hatch. Stick insects are solitary and feed nocturnally. A common situation with stick insects is to have a small shrub partially or completely defoliated overnight. Feeding by related grasshoppers (Orthoptera) can be severe. Lubber and differential grasshoppers are extreme generalists, with relatively few plants that are not hosts. The differential grasshopper is one of the few urban landscape pests that can damage both lawns and ornamental plants (*69*). Differential grasshoppers are widely distributed in the United States, whereas lubber grasshoppers are only common in southern states in coastal areas or along sandy river banks. Both species have one generation per year. Females lay clutches of eggs into soil. The nymphs and adults have plaguelike aggregations when they feed. Although there are no longer swarming locusts in North America, local, very abundant

FIGURE 7.11   Female twostriped walkingstick.

populations of lubbers or differential grasshoppers are reminiscent of swarms. When outbreaks occur, the grasshoppers tend to remain until they have consumed most of the palatable foliage in the area.

Leafcutting bees and ants (Hymenoptera) cut leaves either for food or nest building. Leafcutter ants are only found in subtropical and tropical locations. Leaf-cutter ants are active at night or around dusk, and they may select leaves of plants that are under water stress (70). As leafcutter ants work, they take small sections of leaves into their underground colony and use the leaves to grow fungal gardens. Where they occur, it is common for small woody plants to be defoliated overnight. Leafcutter ants have large numbers of workers in the colony and make large, multilevel mounds. They emerge from the top of the mound, which helps to separate them from imported fire ants. Because they are fungal feeders, leafcutter ants may take leaves from a broad range of plants and do not respond well to most ant baits. Leafcutter bees are elusive, making unique circular cuts to the leaf margin (see Chapter 5). There is no need to control leafcutter bees. Relative to leafcutter ants, the bees harvest relatively small amounts of foliage to make nests. Homeowners or clients with leafcutter bee activity would probably understand when told the holes are caused by beneficial bees that will not defoliate the tree.

As noted for sap-sucking insects and mites, plant identification, type of plant injury (leafminer, defoliator, leaf roll, etc.), and general knowledge of the type of insect that caused that injury (see Chapter 5) are useful knowledge in pest identification. Identification of eruciform larvae to insect family requires knowledge of the host plant, time of year (phenology), and closer examination of the larvae. An insect diagnostic lab can provide an identification; however, the following diagnostic steps can be used to get a more rapid identification of common larvae to family. Because most plant-feeding insects are plant specialists, identification of the host plant can narrow the list of possible species. A useful resource for this step is the HOSTS database *(71)*. This online searchable database lists caterpillars by host plant similar to the aphids database mentioned earlier. Because eruciform larvae are well studied, this database is extensive and reliable. Each plant search will produce a list of candidate larvae, which then requires examination of prolegs and possibly the crochets (see Chapter 5) with a hand lens. Fortunately, most defoliators and the characteristics needed for identification are larger in size than sap suckers. Tables 7.2 and 7.3 lists common families of eruciform larvae with diagnostic features. Among sawfly larvae, the number of prolegs, host plants, and number of antennal segments are important (Table 7.2). Coloration is not always a reliable characteristic for identification of larvae because different color forms of the same species may occur (Fig. 7.12) *(72)* or younger larvae may have different coloration than older larvae.

TABLE 7.2 Diagnostic features useful for family identification of sawfly larvae common in urban landscapes.

Family	Features	Example *(4)*
Diprionidae	Defoliators, uses conifers for larval host plants	White pine sawfly Red-headed sawfly
Tenthridinidae	Broadleaf tree, shrubs, or herbaceous plants Six to eight pairs of prolegs. Most have conical, five segmented antennae	Roseslugs Dusky birch sawfly Alder feeding sawfly *(53)*
Argiidae	Broadleaf tree, shrubs, or herbaceous plants Five to seven pairs of prolegs. One segmented antenna	Hibiscus sawfly
Pamphiliidae	Lack prolegs (atypical for sawflies) Long, seven to eight segmented antennae; noticeable without magnification. Two genera (*Cephalcia* and *Acantholyda*) use pine as hosts	Web-spinning and leafrolling sawflies

For caterpillars of moths and butterflies (Table 7.3), the number of prolegs can be diagnostic. Caterpillars in two families have only two or three pairs of prolegs and often have common names that include the word "looper." Loopers or inch worms have the characteristic walking pattern where they must pull them themselves along with the front legs because they lack additional prolegs to crawl directly like other caterpillars. Crochets on prolegs are arranged in circular, elliptical, or bands patterns. Some families have circular crochet patterns. The individual crochet hooks are also classified as uniordinal (one

FIGURE 7.12 Coloration is not always a reliable characteristic for identification of larvae because different color forms of the same species may occur. For example, the tersa sphinx has brown and green color forms.

TABLE 7.3 Diagnostic features useful for family identification of caterpillars of moths and butterflies common in urban landscapes.

Family	Features	Example
Hesperiidae	Skippers. Larvae have heads wider than the body. Segments with indistinct annulets	Silver-spotted skipper
Papillionidae	Last segment of the thorax usually enlarged. If alive, may evert the osmeterium, a forked defensive organ, near back of head	Swallowtails
Sphingidae	Fleshly caudal horns present in two families: Sphingidae and Bombycidae (silkworms)	Hornworms Catalpa worms
Notodontidae	Prominent larvae; usually have a V shaped hind end with the anal prolegs reduced or missing. Some have a two tone body color and dorsal points on one or more abdominal segments	Azalea caterpillar Walnut caterpillar Oakleaf caterpillar
Arctiinae	Hairs (setae) arranged in tufts on wartlike bumps along the sides and upper part of body. Crochets in a single series	Fall webworm Pecan webworm

Continued

TABLE 7.3 Diagnostic features useful for family identification of caterpillars of moths and butterflies common in urban landscapes.—cont'd

Family	Features	Example
Psychidae	Single larva inside a silken bag. Relatives of clothes moths called plaster bagworms also have one larva per bag but are not plant feeders	Snailcase bagworm Evergreen bagworm
Lasiocampidae	Larvae usually together in silken nest or mat. Do not web over the leaves while feeding. Larvae in this family have an anal point between prolegs	Eastern tent caterpillar Forest tent caterpillar
Lymantriinae	Hairy bodies. Crochets are uniordinal. Some species are stinging caterpillars	Tussock moth Gypsy moth
Galactidae	Larvae web over the leaves on which they feed. Hosts are mimosa or honeylocust. Body is bare with crochets biordinal in a complete circle	Mimosa webworm or honeylocust webworm
Coleophoridae	Larvae feed in individual slender silken tubes that they carry along as they feed	Casebearers, *Coleophora* spp.
Batrachedridae	Palms are the only host. Make tubes of silk covered in frass pellets from which they feed	*Homaledra* spp. Palm leaf skeletonizers or leaf housemakers
"Loopers"	Geometridae if two sets of prolegs and Noctuidae if three sets of prolegs	Cankerworms
Leaf rollers/ leaf tiers	Mainly Torticidae but also Pyralidae. Both families have similar crochet patterns.	Lesser canna leafroller
Stinging caterpillars	Head not visible or hidden. Body has tubercles or fleshy spines or completely covered in hair; some not readily recognizable as larvae. Limacodidae—lack crochets Megalopygidae—hairy "tribble-like" larvae; have crochets Lycaenidae—have crochets	Limacodidae—slug caterpillars Megalopygidae—puss and flannel moths Lycaenidae—hairstreaks

length), biordinal (two lengths), or triordinal (three lengths) based on the different lengths present. The degree of hairiness on caterpillars can also be diagnostic. These hairs found in several families of caterpillars (Table 7.3) can be harmless or they can be stinging or urticating hairs. Urticating hairs may produce a mild prickly sensation on skin or they may connect to sacs containing a poison able to cause intense pain and skin reactions.

The common wood boring beetles and their larvae to family are also distinct enough to identify to family. Roundheaded borer larvae can either be completely legless or have three pairs of legs. The head of roundheaded borers is partially hidden by or retracted into the body producing a rounded appearance to that end of the larva (*4,57,73*). Roundheaded borers develop into longhorn beetles which emerge from the host with a rounded hole, usually much larger than those of the weevils, bark, and ambrosia beetles. In contrast, the larvae of metallic wood—boring beetles (Buprestidae) are called flatheaded borers. Flatheaded borers have one or more enlarged and flattened segments just behind the head (*4,57,73*). Flatheaded borers develop into metallic wood—boring beetles also called jewel beetles. These adults are elliptical and emerge from host plants with a ovoid to D-shaped exit hole (Fig. 7.13). The larvae of bark and ambrosia beetles (Scolytinae) are the smallest (a few millimeters long) of the borers found in ornamentals (*62*). The larvae are only found inside galleries within the tree and appear maggotlike (Fig. 7.10). Caterpillars that bore inside trees resemble leaf-feeding caterpillars. In many cases, legs or prolegs are reduced, but often crochets are still apparent on prolegs. Like borers, miners and larvae of gall-inducing species must be dissected from the plant to be found. The larvae are usually maggotlike and very difficult to identify even to order or family. Many leafminers are host specific and can nearly be identified to species by the appearance of the mine on a particular host plant. In certain situations, plants infested by borers, leafminers, or gall inducers can be held in plastic bins to allow the insects to complete development and emerge as an adult. This will not provide a rapid identification, but the adult life stage that emerges is far easier to identify to species than the larval forms.

FIGURE 7.13 Metallic wood—boring beetles make ovoid- or D-shaped exit holes when the adult beetle emerges.

Highlights

- The species diversity and growth habits of trees, shrubs, and herbaceous annual and perennial plants in urban landscapes provide many unique opportunities (niches) for insects and mites to colonize above ground plant parts.
- Defoliators (leaf chewers), sap suckers, miners, borers, and gall inducers are common niches useful for diagnosing insects and mites on ornamental plants.
- Because many plant-feeding insects are specialists, the identity of the ornamental plant where insects and mites are present is an important step in diagnosing these pests.
- Insects that develop as larvae inside ornamental plants (endophagous) are usually the most difficult to identify and manage.

References

1. Giron, D.; Huguet, E.; Stone, G. N.; Body, M. Insect-induced Effects on Plants and Possible Effectors Used by Galling and Leaf-Mining Insects to Manipulate Their Host-Plant. *Journal of Insect Physiology* **2016,** *84,* 70–89.
2. Kirichenko, N.; Augustin, S.; Kenis, M. J. Invasive Leafminers on Woody Plants: A Global Review of Pathways, Impact, and Management. *Journal of Pest Science* **2019,** *92,* 93–106.
3. Liu, H. Under Siege: Ash Management in the Wake of the Emerald Ash Borer. *Journal of Integrated Pest Management* **2018,** *9,* 1–16.
4. Johnson, W. T.; Lyon, H. H. *Insects That Feed on Trees and Shrubs,* 2nd ed.; Cornell University Press: Ithaca, 1991.
5. Braman, S. K.; Buntin, G. D.; Oetting, R. D. Species and Cultivar Influences on Infestation by and Parasitism of a Columbine Leafminer (*Phytomyza aquilegivora* Spencer). *Journal of Environmental Horticulture* **2005,** *23* (1), 9–13.
6. Mani, M. S. *Ecology of Plant Galls;* Springer: Netherlands: The Hague, 1964.
7. Richardson, R. A.; Body, M.; Warmund, M. R.; Schultz, J. C.; Appel, H. M. Morphometric Analysis of Young Petiole Galls on the Narrow-Leaf Cottonwood, *Populus angustifolia,* by the Sugarbeet Root Aphid, *Pemphigus betae. Protoplasma* **2017,** *254,* 203–216.
8. Raupp, M. J.; Buckelew, A.; Raupp, E. C. Street Tree Diversity in Eastern North America and It's Potential for Tree Loss to Exotic Borers. *Arboriculture & Urban Forestry* **2006,** *32* (6), 297–304.
9. Raupp, M. J.; Shrewsbury, P. M.; Holmes, J. J.; Davidson, J. A. Plant Species Diversity and Abundance Affects the Number of Arthropod Pests in Residential Landscapes. *Arboriculture & Urban Forestry* **2001,** *27* (4), 222–229.
10. Stewart, C. D.; Braman, S. K.; Sparks, B. L.; Williams-Woodward, J. L.; Wade, G. L.; Latimer, J. G. Comparing an IPM Pilot Program to a Traditional Cover Spray Program in Commercial Landscapes. *Journal of Economic Entomology* **2002,** *95* (4), 789–796.
11. Eliason, E. A.; Potter, D. A. Biology and Management of the Horned Oak Gall Wasp on Pin Oak. *Journal of Arboriculture* **2001,** *27,* 92–101.
12. Sacchi, C. F.; Connor, E. F. Changes in Reproduction and Architecture in Flowering Dogwood, *Cornus florida,* after Attack by the Dogwood Club Gall, *Resseliella clavula. Oikos* **1999,** *86* (1), 138–146.
13. Felt, E. P. *Plant Galls and Gall Makers;* Hafner Press: New York, 1965.

14. Gagné, R. J. *The Plant-Feeding Gall Midges of North America;* Cornell University Press: Ithaca, 1989.

15. Hodges, A.; Buss, E.; Mizell, R. F., III. *Insect Galls of Florida;* University of Florida: Gainesville, 2006.

16. Tooker, J. F.; Rohr, J. R.; Abrahamson, W. G.; De Moraes, C. M. Gall Insects Can Avoid and Alter Indirect Plant Defenses. *New Phytologist* **2008,** *178,* 657–671.

17. Yamazaki, K. Caterpillar Mimicry by Plant Galls as a Visual Defense against Herbivores. *Journal of Theoretical Biology* **2016,** *404,* 10–14.

18. Eliason, E. A.; Potter, D. A. Dogwood Borer (Lepidoptera: Sesiidae) Infestation of Horned Oak Galls. *Journal of Economic Entomology* **2000,** *93,* 757–762.

19. Cranshaw, W.; Shetlar, D. *Garden Insects of North America: The Ultimate Guide to Backyard Bugs,* 2nd ed.; Princeton University Press: Princeton, 2017.

20. Miller, D. R.; Miller, G. L.; Hodges, G. S.; Davidson, J. A. Introduce Scale Insects (Hemiptera: Coccoidea) of the United States and Their Impact on U.S. Agriculture. *Proceedings of the Entomological Society of Washington* **2005,** *107* (1), 123–158.

21. Doo-Hyung, L.; Yong-Lak, P.; Leskey, T. C. A Review of Biology and Management of *Lycorma delicatula* (Hemiptera: Fulgoridae), an Emerging Global Invasive Species. *Journal of Asia-Pacific Entomology* **2019,** *22* (2), 589–596.

22. Panfilio, K. A.; Angelini, D. R. By Land, Air, and Sea: Hemipteran Diversity through the Genomic Lens. *Current Opinion in Insect Science* **2018,** *25,* 106–115.

23. Moritz, G. Structure, Growth and Development. In *Thrips as Crop Pests;* Lewis, T., Ed.; CAB: New York, 1997; pp 15–63.

24. Bensoussan, N.; Santamaria, M. E.; Zhurov, V.; Diaz, I.; Grbic, M.; Grbic, V. Plant-herbivore Interaction: Dissection of the Cellular Pattern of *Tetranychus urticae* Feeding on the Host Plant. *Frontiers of Plant Science* **2016,** *7,* 1105.

25. Nahrung, H. F.; Waugh, R. Eriophyid Mites on Spotted Gums: Population and Histological Damage Studies of an Emerging Pest. *International Journal of Acarology* **2012,** *38,* 549–556.

26. Park, Y. L.; Lee, J. H. Leaf Cell and Tissue Damage of Cucumber Caused by Twospotted Spider Mite (Acari: Tetranychidae). *Journal of Economic Entomology* **2002,** *95* (5), 952–957.

27. Young, R. F.; Shields, K. S.; Berlyn, G. P. Hemlock Wooly Adelgid (Homoptera: Adelgidae): Stylet Bundle Insertion and Feeding Sites. *Annals of the Entomological Society of America* **1995,** *88* (6), 827–835.

28. Chen, Y.; Yang, C.; Holford, P.; Beattie, G. A.; Spooner-Hart, R. N.; Liang, G.; Deng, X. Feeding Behavior of the Asian Citrus Psyllid, *Diaphorina citri,* on Healthy and Huanglongbing-Infected Citrus. *Entomologia Experimentalis et Applicata* **2012,** *143,* 13–22.

29. Walstad, J. D. Effect of the Pine Needle Scale on Photosynthesis of Scots Pine. *Forest Science* **1973,** *19* (2), 109–111.

30. Sadof, C. S.; Neal, J. J. Use of Host Plant Resources by the Euonymus Scale, *Unaspis euonymi* (Homoptera: Diaspididae). *Annals of the Entomological Society of America* **1993,** *86* (5), 614–620.

31. Hanks, L. M.; Denno, R. F. The White Peach Scale, *Pseudaulacaspis pentagona* (Targioni: Tossetti) (Homoptera: Diaspididae): Life History in Maryland, Host Plants and Natural Enemies. *Proceedings of the Entomological Society of Washington* **1993,** *95,* 79–98.

32. Buntin, G. D.; Braman, S. K.; Gilbertz, D. A.; Phillips, D. V. Chlorosis, Photosynthesis, and Transpiration of Azalea Leaves after Azalea Lace Bug (Heteroptera: Tingidae) Feeding Injury. *Journal of Economic Entomology* **1996,** *89* (4), 990–995.

33. Backus, E. A.; Serrano, M. S.; Ranger, C. M. Mechanism of Hopperburn: An Overview of Insect Taxonomy, Behaviors, and Physiology. *Annual Review of Entomology* **2005,** *50,* 125—151.

34. Lucini, T.; Panizzi, A. R. Electropenetrography (EPG): A Breakthrough Took Unveiling Stink Bug (Pentatomidae) Feeding on Plants. *Neotropical Entomology* **2018,** *47,* 6—18.

35. Cockfield, S. D.; Potter, D. A. Chlorosis and Reduced Photosynthetic CO_2 Assimilation of *Euonymus fortunei* Infested with Euonymus Scale (Homoptera: Diaspididae). *Environmental Entomology* **1987,** *16* (6), 1314—1318.

36. Sicard, A.; Zeilinger, A. R.; Vanhove, M.; Schartel, T. E.; Beal, D. J.; Daugherty, M. P.; Almeida, R. P. P. *Xylella fastidiosa*: Insights into an Emerging Plant Pathogen. *Annual Review of Phytopathology* **2018,** *56,* 181—202.

37. Alverson, D. R.; Allen, R. K. *Life History of the Crapemyrtle Aphid.* Proceedings of the Southern Nursery Association Research Conference **1991,** *36*; pp 164—167.

38. Killiny, N.; Jones, S. E. Metabolic Alterations in the Nymphal Instars of *Diaphorina Citri* Induced by *Candidatus Liberibacter asiaticus*, the Putative Pathogen of Huanglongbing. *PLoS One* **2018,** *13* (1), e0191871.

39. Brodbeck, B. V.; Andersen, P. C.; Mizell, R. F. Differential Utilization of Nutrients During Development by the Xylophagous Leafhopper, *Homalodisca coagulata*. *Entomologia Experimentals et Applicata* **1995,** *75,* 279—289.

40. Cockfield, S. D.; Potter, D. A. Euonymus Scale (Homoptera: Diaspididae) Effects on Plant Growth and Leaf Abscission and Implication for Differential Site Selection by Male and Female Scales. *Journal of Economic Entomology* **1990,** *83* (3), 995—1001.

41. Gullan, P. J.; Kosztarab, M. Adaptations in Scale Insects. *Annual Review of Entomology* **1997,** *42,* 23—50.

42. Glynn, C.; Herms, D. A. Local Adaptation in Pine Needle Scale (*Chionaspis pinifoliae*): Ntal and Novel Host Quality as Tests for Specialization Within and Among Red and Scots Pine. *Environmental Entomology* **2004,** *33* (3), 748—755.

43. Magsig-Castillo, J.; Morse, J. G.; Walker, G. P.; Bi, L. B.; Rugman-Jones, P. F.; Stouthamer, R. Phoretic Dispersal of Armored Scale Crawlers (Hemiptera: Diaspididae). *Journal of Economic Entomology* **2010,** *103* (4), 1172—1179.

44. Koteja, J. Life History. In *The Armored Scale Insects, their Biology, Natural Enemies, and Control;* Rosen, D., Ed.; Elsevier Science Publishing: Amsterdam, The Netherlands, 1990; pp 243—254.

45. Hubbard, J. L.; Potter, D. A. Life History and Natural Enemy Associations of Calico Scale (Homoptera: Coccidae) in Kentucky. *Journal of Economic Entomology* **2005,** *98* (4), 1202—1212.

46. Pemberton, H. B.; Ong, K.; Windham, M.; Byrne, D. H. What Is Rose Rosette Disease? *HortScience* **2018,** *53* (5), 592—595.

47. Hodges, A.; Ludwig, S.; Osborne, L.; Edwards, G.B. Pest Thrips of the United States: Field Identification Guide. https://www.ncipmc.org/action/chili_thrips_deck.pdf.

48. Aphids of the World Database. http://www.aphidsonworldsplants.info/C_HOSTS_AAIntro.htm.

49. Hamon, A. B.; Williams, M. L. *Arthropods of Florida and Neighboring Land Areas Vol. 11 the Soft Scale Insects of Florida (Homoptera: Coccoidea: Coccidae)*, 1984; p 194. Available online at: https://palmm.digital.flvc.org/islandora/object/uf%3A46687#page/i/mode/2up.

50. Dekle, G. W. *Arthropods of Florida and Neighboring Land Areas Vol. 3 Florida Armored Scale Insects,* 1965; p 265. Available online at: http://palmm.digital.flvc.org/islandora/object/uf%3A97189#page/dpi/mode/2up.

51. Hodges, G.S.; Evans, G.A. Key to the families of Scale Insects in Florida (Adult Females) http://www.fsca-dpi.org/Homoptera_Hemiptera/scales/Scale_Families.pdf.

52. Hodges, A.; Hodges, G.; Buss, L.; Osborne, L. Mealybugs & Mealybug Look-alikes of the Southeastern United States. https://www.ncipmc.org/action/alerts/mealybugs.pdf.

53. Looney, C.; Smith, D. R.; Collman, S. J.; Langor, D. W.; Peterson, M. A. Sawflies (Hymenoptera: Symphyta) Newly Recorded from Washington State. *Journal of Hymenoptera Research* **2016,** *49,* 129−159.

54. Berenbaum, M. Coumarins and Caterpillars: A Case for Coevolution. *Evolution* **1983,** *37,* 163−179.

55. Taft, W. H.; Smitley, D.; Snow, J. W. *A Guide to the Clearwing Borers (Sesiidae) of the North Central United States;* North Central Regional Publication, 1994; p 30. No.394.

56. Held, D. W.; Pickens, J. *Borer Pests of Woody Ornamental Pests. ANR-2472;* Alabama Cooperative Extension System, 2018; p 10.

57. Solomon, J.D. Guide to Insect Borers in North American Broadleaf Trees and Shrubs. United States Department of Agriculture. Forest Service Agriculture Handbook AH-706. (Available online at: https://www.srs.fs.usda.gov/pubs/misc/ah_706/ah-706.htm.

58. Rogers, L. E.; Grant, J. E. Infestation Levels of Dogwood Borer (Lepidoptera: Sesiidae) Larvae on Dogwood Trees in Selected Habitats in Tennessee. *Journal of Entomological Science* **1990,** *25* (3), 481−485.

59. Weissling, T.J.; Giblin-Davis, R.M. Palemetto weevil, EENY-13. http://entnemdept.ufl.edu/creatures/orn/palmetto_weevil.htm.

60. Dembilio, Ó.; Jacas, J. A.; Llácer, E. Are the Palms *Washingtonia filifera* and *Chamaerops humilis* Suitable Hosts for the Red Palm Weevil, *Rhynchophorus Ferrugineus* (Col. Curculionidae)? *Journal of Applied Entomology* **2009,** *133,* 565−567.

61. Beetles, B. *Biology and Ecology of Native and Invasive Species;* Elsevier/Academic Press: London, 2015.

62. Ranger, C. M.; REding, M. E.; Schultz, P. B.; Oliver, J. B.; Frank, S. D.; Adesso, K. M.; Chong, J. H.; Sampson, B.; Werle, C.; Gill, S.; Krause, C. Biology, Ecology, and Management of Nonnative Ambrosia Beetles (Coleoptera: Curculionidae: Scolytinae) in Ornamental Plant Nurseries. *Journal of Integrated Pest Management* **2016,** *7* (1), 1−23, 9.

63. Dute, R. R.; Miller, M. E.; Davis, M. A.; Woods, F. M.; McLean, K. S. Effects of Ambrosia Beetle Attack on *Cercis canadensis. International Association of Wood Anatomists Journal* **2002,** *23,* 143−160.

64. Mizell, R.; Riddle, T. C. Evaluation of Insecticides to Control the Asian Ambrosia Beetle, Xylosandrus crassiusculus. *Proceedings of the Southern Nursery Association Research Conference* **2004,** *49,* 152−155.

65. Haack, R. A. Feeding Biology of Cerambycids. In *Cerambycidae of the World; Biology and Pest Management;* Wang, Q., Ed.; CRC Press: Boca Raton, Florida, 2017; pp 105−124.

66. Muilenburg, V. L.; Herms, D. A. A Review of the Bronze Birch Borer (Coleoptera: Buprestidae) Life History, Ecology, and Management. *Environmental Entomology* **2012,** *41* (6), 1372−1385.

67. Korotyaev, B. A.; Konstantinov, A. S.; Lingafelter, S. W.; Mandelshtam, M. Y.; Volkovitsh, M. G. Gall-inducing Coleoptera. In *Biology, Ecology, and Evolution of Gall-Inducing Arthropods;* Raman, A., Schaefer, C. W., Withers, T. M., Eds.; Science Publishers: Plymouth, United Kingdom, 2005; pp 239−271.

68. Condra, J. M.; Brady, C. M.; Potter, D. A. Resistance of Landscape-Suitable Elms to Japanese Beetle, Gall Aphids, and Leaf Miners, with Notes Life History of *Orchestes alni* and *Agromyza aristata* in Kentucky. *Arboriculture & Urban Forestry* **2010,** *36* (3), 101−109.

69. Reinert, J. A.; Mackay, W.; Englke, M. C.; George, S. W. The Differential Grasshopper (Orthoptera: Acrididae)-Its Impact on Turfgrass and Landscape Plants in Urban Environs. *Florida Entomologist* **2011**, *94* (2), 253–261.

70. Meyer, S. T.; Roces, F.; Wirth, R. Selecting the Drought Stressed: Effects of Plant Stress on Intraspecific and Within-Plant Herbivory Patterns of the Leaf-Cutting Ant *Atta colombica. Functional Ecology* **2006**, *20,* 973–981.

71. Robinson, G.S.; Ackery, P.R.; Kitching, I.J.; Beccaloni, G.W., Hernandez, L.M. HOSTS—a database of the world's Lepidopteran Hostplants. Natural History Museum http://www.nhm.ac.uk/research-curation/research/projects/hostplants/.

72. Schowalter, T. D.; Ring, D. R. Biology and Management of the Fall Webworm, *Hyphantria cunea* (Lepidoptera: Erebidae). *Journal of Integrated Pest Management* **2017,** *8* (1), 1–6, 7.

73. Nearns, E.H.; Redford, A.J.; Walters, T.; Miller, K.B. A Resource for Wood Boring Beetles of the World. The University of New Mexico and Center for Plant Health Science and Technology, USDA, APHIS, PPQ. Available from: http://wbbresource.org/.

Chapter 8

Nonchemical approaches to pest management

Introduction

Nonchemical approaches in pest management are an important component of integrated pest management (IPM) in urban landscapes. The concept of IPM, introduced in Chapter 5, uses knowledge of both pest and plant biology to reduce or prevent damage and loss using methods that are consistent with stewardship of the environmental and our natural resources. Someone unfamiliar with IPM may ask the question, "Why not just use insecticides?" Insecticides are fast acting and relatively easy to use, but they are not a good option for every situation, nor do they provide long-term management options. While insecticides are short-term, reactive interventions to reduce populations of pest insects and mites, nonchemical approaches are preventive or proactive ways to interrupt pest biology or behaviors. This chapter will define and discuss biological and cultural approaches used in urban landscapes.

Biological controls

Natural enemies of pests can either be predators, parasitoids, microbes, or nematodes. Predators consume their prey, causing prompt death. Ants, certain true bugs and beetles, and spiders are the main groups of predators. Larger animals, mice, skunks, armadillos, and birds are also common predators in urban landscapes. An adult parasitoid would attack a pest, but the pest is later killed by the larval stage. Larvae of parasitoids develop internally or externally slowly consuming a host over time. Parasitoids are further classified by the life stage on which the larvae develop. *Tiphia* wasps, for example, are larval parasitoids that lay eggs on white grubs.

Biological controls and natural enemies have been mentioned throughout this book. These terms are commonly used interchangeably to mean the animals and microbes, mostly endemic in urban landscapes, which help reduce pest populations and the frequency of outbreaks *(1)*. However, this definition best

Urban Landscape Entomology. https://doi.org/10.1016/B978-0-12-813071-1.00008-7

defines natural enemies with biological control being the intentional manipulation of natural enemies as part of IPM. Natural enemies can cause significant mortality of landscape pests, particularly scale insects on trees and shrubs
(2), and white grubs and caterpillars in turfgrass (3,4). However, mortality
from natural enemies may not be consistent or significantly reduce damage to
an acceptable esthetic level.

There are three ways, Classical, Conservation, and Augmentation, to
introduce biological control agents or to improve mortality from existing
natural enemies (5). Classical biological control is where natural enemies of
exotic pests are identified in their native geographic range. Those natural
enemies are then imported and released in new locations where the pest has
been introduced. The history of classical biological control in North America
began when an expedition commission under the direction of federal entomologist C.V. Riley went to Australia and found natural enemies of the cottony
cushion scale (6). Since then, hundreds of parasitoids, predators, and nematodes have been imported and released to reduce populations of insects that
attack woody plants and turfgrass. This is done under the regulation of the US
Department of Agriculture, and it is not a practice where homeowners or
businesses can order exotic natural enemies online. Each imported biological
control agent is evaluated under quarantine to avoid unintended mortality of
native species. This biological control practice is usually reserved for the most
damaging pest species such as eucalyptus longhorned borer, ash whitefly,
Japanese beetles, mole crickets, and others.

Endemic natural enemies or classical biological control agents may need
assistance to maintain or enhance their populations. This is a practice of
Conservation biological control. The two main research areas in Conservation
biological control are (1) the compatibility of insecticides with natural enemies
(discussed in Chapter 9) and (2) providing habitat or food for natural enemies
to increase populations locally (7). Conservation biological control is based on
well-documented relationships between survival of natural enemies and lifetime predation or parasitism. Basically, providing food or other hosts attracts,
maintains, or prolongs the life of parasitoids or predators, which in turn allows
them to impose greater reductions on pest populations. In an earlier chapter,
the influence of structural complexity on natural enemies and pest populations
was discussed. While it is not feasible to instantly add overstory trees in the
short term, increasing vegetation in each of the layers increases complexity
and makes the landscape better suited for recruiting and retaining natural
enemies. Several published studies on Conservation biological control in urban
landscapes focus on specific planting for habitat (beetle banks) or nectar
sources for natural enemies (8–15). The flowering of these plants should
coincide with the activity of important natural enemies. Plantings can serve a
dual role in providing shelter and food for natural enemies. Rebek et al. (12)
removed the inflorescences from plants with no difference in abundance of
natural enemies relative to plots with flowers intact. *Larra bicolor,* parasitoid

of mole crickets, commonly rest on nectar sources during the day and may remain overnight on their nectar source plants *(14)*. There are few plant species that attract or supplement natural enemies in all situations. Buckwheat (*Fagopyrum*), goldenrod (*Solidago*), shasta daisy (*Leucanthemum*), aster (*Aster*), and tickseed (*Coreopsis*) are floral resources commonly used in experiments. These plant families Apiaceae, Anacardiaceae, Asteraceae (aster, coreopsis, and shasta daisy), Polygonaceae (buckwheat), Rhamnaceae, Rutaceae, and Salicaceae have also been suggested to host high numbers of parasitoids *(16)*. If flowers that support parasitoids of a particular pest are not known, then plants in the family Apiaceae, which have exposed nectaries, would be a good starting point *(16)*.

Conservation biological control is also accomplished by introducing food supplements or volatile attractants that are not food. Birds are significant predators of insects, particularly larger caterpillars, in woody plants. The presence of birds can be directly related to a reduced presence of caterpillars, reduced leaf damage, or increased growth due to caterpillar predation *(17−22)*. The effects of bird predation also should be greater in inner city areas compared with rural areas *(22)*. When birds are excluded experimentally, caterpillar density and proportion of branches with caterpillars are 24% greater than trees where birds had access to caterpillars *(20)*. Based on this, recreational bird feeding can aggregate birds in landscapes and increase predation on adjacent trees up to 20 m (65.6 ft) from the feeder or food source *(23)*. It is not just caterpillar predation. Providing seeds for smaller songbirds (passerine birds) in gardens can reduce the abundance and the duration of pea aphids on beans placed in the gardens *(24)*. Bird feeding, however, may not be selective to pests. Small ground beetles (predators) that may be attracted to seeds near bird feeders are also more likely to be consumed by birds *(25)*. For insects, the common nonplant supplements are food sprays such as synthetic honeydew or sugar water *(10)*. These food supplements are applied to, or adjacent to, infested plants. Sugar sprays have been evaluated with parasitoids of white grubs and mole crickets. Compounds such as methyl salicylate, jasomnates, and jasmonic acid, are being experimentally evaluated as attractants for natural enemies in search of prey or hosts. These compounds are host-induced plant volatiles or plant signaling chemicals, which exploit the inherent attraction of natural enemies to plants infested with herbivores *(26)*. While research studies exist in food crops *(27)*, there are few studies evaluating these compounds with pests in urban landscapes.

Augmentation is the third practice of biological control. Like classical biological control, augmentation also releases natural enemies, except these native or endemic natural enemies can be purchased from vendors of beneficial organisms *(28)* and released locally. Most companies marketing natural enemies will assist their customers with the correct choice of product and release support, or you could consult one of the general books on biological control for suggestions *(5)*. Augmentation happens by either inoculation or inundation. Inoculation is the release of natural enemies where the goal is not an immediate suppression of the pest population. Inoculation is the introduction and gradual increase in populations over time. Inoculation, for example, would be the fixed release of 100 parasitoids of whiteflies monthly. Inoculation aims to produce a more constant pressure of natural enemies on pests. It is not used for a rapid response to a pest outbreak. Inundation is the release of natural enemies with the goal of more rapid pest suppression. Because immediate effects are needed, inundation uses predatory insects or mites because they kill their prey immediately. Of these two release strategies, there are only studies using inundation in urban landscapes, outdoor production of ornamentals, or turfgrass (29–36). Among these studies, uses of entomopathogenic nematodes and microbes are primarily targeting soil-dwelling insects, and use of arthropods is primarily targeting insect and mite pests of woody ornamentals.

Nematodes, viruses, fungi, or bacteria can be formulated into products that can be applied in water using equipment used to apply pesticides. Pathogens have specific routes by which they attack pests, but all must get inside the insect to be effective. Fungi grow on the surface of the insect getting inside by exploiting breaks or thin areas in the outermost cuticle layer of the insect exoskeleton. Viruses and bacteria are consumed inadvertently during feeding, and entomopathogenic nematodes actively invade through natural openings like spiracles. Grewal (37) provides a summary of pathogens produced commercially for use against turfgrass pests. A far greater number of pathogens are reported from insects, but a smaller number of those have been formulated into products. For example, fall armyworms are susceptible to at least 20 pathogens (38), but few microbial insecticides have become products (37). There are limitations on producing biological materials that are shelf stable and that meet a customer's expectation to perform like insecticides (39). One viral pathogen of black cutworms has been well studied (40,41) and eventually may become the first viral insecticide used in turfgrass. This virus can be sprayed onto fairway height creeping bentgrass and provides ≥75% mortality of larvae that encounter fresh sprays. The virus continues to provide about 50% mortality of black cutworm larvae even 28 days after being sprayed. Once infected by the virus, the caterpillars reduce food consumption and die within 6–9 days. Following death, the body liquifies releasing virus particles (virions) into the environment where they may persist over weeks or years (40). Fungal insect pathogens applied to turfgrass are either *Beauveria* spp. or *Metarhizium anisopliae*. In controlled lab and greenhouse experiments, fungal pathogens can reduce survival of turfgrass insects, particularly chinch bugs and mole crickets (42–45). *Beauveria bassiana* and other fungal pathogens may also convey resistance to white grubs through colonization of roots as endophytes (46). However, the effectiveness is less apparent in field tests. *Beauveria bassiana* applied to bermudagrass in July or August has the greatest viability (70%) 1 day after application, after which it declines to less than 50%

over 3—7 days postapplication. More frequent irrigation can slightly increase viability by 10%, but viability still exhibits a steep decline 1 day after application *(44)*. This may explain why field applications of the same pathogen failed to reduce damage from mole crickets *(47)*.

Bacteria and entomopathogenic nematodes have the best records of development into products for use against turfgrass pests. *Bacillus thuringiensis* (Bt) and *Paenibacillus popilliae* and *Paenibacillus lentimorbus* (milky disease) *(48)* are marketed as alternatives to conventional insecticides for caterpillars and white grubs. In fact, Rachael Carson, author of *Silent Spring (49)*, praised the innovation and selectivity of these two bacteria in her book. Both bacteria are naturally occurring in soil but have been formulated into products likely on the shelf of your local garden center or box store. These products were developed to add to, or augment, local populations to increase disease incidence. For Bt, specific subspecies and strains (analogous to bacterial cultivars) provide target-specific control. For example, Bt *buibui* and *galleriae* have activity against white grubs, *kurstaki* has activity against caterpillars, and *israeliensis* has activity against mosquito larvae and European crane fly *(39)*. Bt *buibui* activity against Japanese beetle and oriental beetle grubs can be 60%—70% when timed against smaller (second instar) grubs, but similar treatments failed to effectively control either Asiatic garden beetle or European chafer grubs *(46)*. Milky disease is often detected at low levels ($\leq$19% infection rate) in field surveys of white grubs *(50—52)*. Field tests attempting to augment local populations of *Paenibacillus* bacteria often fail to show an increasein disease incidence in white grub populations *(53)* or similar effectiveness as insecticides *(54)*. *Serratia entomophila* infection of white grubs is associated with a condition called amber disease. Isolates of this bacterium are incidentally collected in surveys of white grubs, but the incidence is usually very low ($\leq$10%) *(50,51)*. It appears to be present in the United States, but the role in reducing white grub populations appears minimal.

Similar to the other microbes, the detection of soil insects infected with entomopathogenic nematodes (EPNs) in the field is relatively low *(51,52)*, and few studies have surveyed the natural occurrence of EPN in populations of aboveground turfgrass pests. Much of the published research on EPN in turfgrass concerns the application of a few species within two genera (*Steinernema* and *Heterorhabditis* spp.) to turfgrass for biological control. In pest management, EPNs are applied in the infective juvenile stage at 1—2.5 billion EPNs per ha (0.4—1 billion EPNs per acre) or 1—10 million per liter. They are formulated to be applied using common spray equipment. After application, turf is usually watered for applications targeting above- or belowground pests *(55,56)*. Infective juveniles seek insect hosts to infect but still must successfully enter the insect through spiracles or other openings. Upon entering, they release bacteria that kill the insect in 1—3 days. The nematodes then reproduces inside the insect, generally filling the entire body with new nematodes. Eventually, the body wall bursts open releasing those new infective juveniles into the environment. Field studies evaluating EPNs in turfgrass are available for all major pests in turfgrass except ground pearls, with the majority of studies concerning turfgrass insects in soil. Among studies with EPNs, four common outcomes are noted. (1) **Persistence**: After application, EPNs can persist for years providing suppression but often not complete control of the target insects *(56,57)*. Persistence just means that infected insects were detected months or years after EPNs were applied. Often there is a decline in EPN populations within weeks or a few months after application *(35,56,57)*. (2) **Compatibility and synergy with certain insecticides**: *Steinernema scapterisci*, a nematode that targets mole crickets, can remain viable even if mixed with many common turfgrass insecticides *(58)*. Furthermore, neonicotinoid and perhaps other insecticides may induce sluggish behavior in white grubs that enhances the ability of nematodes to enter the insect *(59)*. (3) **High cost and limited availability:** The use rates require companies to produce large numbers of living organisms and have them ready "on the shelf" to fit the market. Furthermore, it can be difficult to judge the market demand from year to year. In 2006, Barbara and Buss *(57)* noted persistence of *S. scapterisci* for about 12 years in managed turfgrass. In the few years following that report, it was difficult to find those nematodes commercially due to higher demand. It is not often the choice economically to use EPNs. Insecticides are often far less expensive than nematodes for control of turfgrass pests. This likely explains why microbials account for less such a small share of the overall turfgrass insecticide market *(37)*. (4) **EPNs may not reduce insect populations below what is aesthetically acceptable.** Admittedly, there are less data in support of this than the other outcomes. Most lab and field studies are primarily concerned with application and infection rates. However, infection rates may not be an indication of turfgrass quality or damage *(57)*.

Lacewings (Neuroptera: *Chrysoperla* spp.), predatory mites, and different species of lady beetles are studied in the published work, targeting either pest aphids (by lacewings and lady beetles) or spider mites (by predatory mites). Raupp *(29)* presents the results of three field experiments with aphids on different ornamentals. The release of convergent lady beetles reduced the percent of infested shoots by 47%, with a 77% decrease in aphids per eleagnus leaf 15 d after the released. The release of lacewing larvae to control aphids on the perennial stonecrop or Washington hawthorn trees did not decrease aphids after release *(29)*. Convergent lady beetles released at 2300 beetles per m^2 of rose hedge provide 93%−100% control of rose aphids *(31)*. They further calculated the effective release rates based on the typical shipping volume of lady beetles. A 0.5-L shipment would treat six landscape roses, 0.5−1 m tall, and a 3.8-L (1 gallon) size shipment of lady beetles would treat 51 landscape roses *(31)*. Few other augmentation experiments or case studies conducted under field conditions are available.

The application of biological control as part of IPM in urban landscapes requires additional discussion about the pre-existing conditions, expectations, and costs. First, the existing level of live or parasitized insects, particularly for small insects (aphids, whiteflies, scale insects), can help determine the next steps. Parasitism will appear as different symptoms in different species (Fig. 8.1) Although it is not possible to catalog the different appearances of parasitized insects here, signs and symptoms of parasitism are available online or from other resources (5). Once familiar with the appearance, it is easy to estimate the total percentage of dead insects. Take a sample of three to five shoots or leaves per plant and examine 25—50 insects, and then express the dead insects as a percent. If the percentage of dead scales is high (>50%), you may opt to use conservation biological control to supplement the natural enemies already present. If the percent of dead insects is low, reduced-risk insecticides (see Chapter 9) or augmentative release of predators may be appropriate. Whatever form of biological control is decided, it is important to not overestimate the potential impacts. In the published examples, the amount of control varied widely with the type of biological control practice used. Conservation biological control yielded low (0%—50%) reductions in pest populations (9,60), where augmentation can provide 50%—100% reduction. Landscape managers should consider the level of control expected if biological control is considered. For example, in a public park, 24%—50% reduction in scale populations from using annual or perennials plants may be sufficient. However, biological control through augmentation, which more effectively and quickly reduces populations, may be required for highly maintained residential landscapes. Conservation biological control can also have a limited range of effectiveness. Conservation strips are narrow plantings of flowering

FIGURE 8.1 An aggregation of aphids showing the healthy individuals with the reddish coloration and the parasitized "mummy" that has the yellowish-brown appearance.

perennials. When installed adjacent to fairways on a golf course, there is a local increase in abundance of predators and parasitoids. Increased predation of turf-infesting caterpillars was limited to within 4 m of the conservation strip *(9)*. The close proximity of the floral resources is consistent with studies with parasitoids. Parasitism usually decreases with increasing distance from the nectar source *(10,11,61)*. Floral resources for natural enemies also risk providing resources for some nectar-feeding adult stages of pests (beetles and moths). The presence of nectar-producing flowers may attract moths and increase oviposition locally *(62,63)* if the nectar sources are generic. These negative effects of nectar sources would be most evident for moths with multiple generations per year such as turfgrass-infesting caterpillars. Ants are another group of insects noted to interfere with the practices of biological control. Ants are reported to protect honeydew-producing insects from parasitoids *(29,64)*, and some biological control programs must selectively remove or exclude ants to be successful *(32,65)*.

As noted for EPNs, augmentation biological control with arthropods can be expensive relative to insecticides. In the previously mentioned experiment with roses and rose aphids *(31)*, three releases of lady beetles were needed to maintain low populations of aphids. Lady beetles consume about 100 aphids in the first 3 days after release and then leave without necessarily laying eggs *(32)*. Since aphids have multiple generations per year, multiple releases are needed. At the time this research was published, each release costs $1.30–$7.20 per landscape rose using the 2300 beetles per m^2 release rate. Three releases of lady beetles are similar to or slightly greater than the per plant costs of applying a systemic insecticide *(31)*. Other examples in the literature suggest costs of augmentative biological control far exceed (2.5 to 7 times) the costs of insecticide applications *(33)*. At the currently documented levels of increased parasitism (25% greater) from sugar or food sprays, it seems unlikely that landscape managers would invest labor to treat with a food spray for such a low rate of return.

Cultural controls

Cultural controls are interventions to reduce pest populations that exploit host plant resistance, and the behavior and ecology of pests (Tables 8.1 and 8.2). Host plant resistance is based on observations or experimental evaluation of plants species or cultivars for susceptibility to one or more pests. These data are then used to advise horticulturalists, urban foresters, and landscape architects on plants that will require less or no insecticides to manage an insect or mite pest. Host plant resistance is defined broadly as resistance and tolerance *(66,67)*. Tolerance describes the ability of a plant to grow in spite of insect herbivory, and resistance is a plant trait(s) that reduces or prevents injury from an herbivorous insect or mite. There are mechanisms underlying what is perceived as host plant resistance. Two mechanisms, antibiosis and antixenosis, were originally recognized by Painter *(66)* but have since been expanded to four mechanisms: constitutive, induced, direct, and indirect defenses. Constitutive defenses are present without prior damage from herbivores, with induced defenses only expressed in response to an herbivore attack. An example of induced defenses would be poor performance of a caterpillar feeding on a plant already damaged by another leaf-feeding insect earlier in the year. Induced plant defenses are associated with one of two plant signal pathways: the jasmonic acid or salicylic acid pathways. The salicylic acid pathway is induced by sap-sucking insects, plant viruses, and certain plant pathogens such as rust or powdery mildew (biotrophic pathogens). The jasmonic acid pathway is induced by chewing insects, plant parasitic nematodes,

TABLE 8.1 Cultural control options for selected ornamental pests.

Pest	Cultural control
Eastern tent caterpillar	Remove small nests or egg masses
Bagworms	Remove bags and destroy in the winter or before egg hatch *(94)* Resistant maple species and cultivars *(95)*
Canna leafroller	Resistant cultivars *(96)* Remove infested leaves or overwintering pupae in plant stubble
Lantana lace bug	Resistant cultivars *(97)*
Japanese beetle	Resistant cultivars and species of woody plants *(98,99)* Resistant cultivars of cannas *(96)* Remove rose blooms *(85,86)* Hand remove beetles at temperatures lower than 23°C (73°F) *(100,101)*.
Azalea lace bug	Resistant cultivars (reviewed in *(102)*)
Western flower thrips	Resistant cultivars of rose *(103,104)*, chrysanthemum *(105)*, and impatiens *(106)*
Soft scale insects	Dislodge with water washing *(107)*
Gall aphids	Resistant elm cultivars *(99)*
Aphids	Dislodged with a strong stream of water *(108)*
Leafminers	Resistant elm cultivars *(99)* Resistant columbine cultivars *(109)*
Twig girdlers and pruners	Pick up and discard any limbs or twigs under large trees *(84)*.

TABLE 8.2 Cultural control options for selected turfgrass pests.

Pest	Cultural control
White grubs	Reduce or eliminate lighting to prevent damage from sugarcane beetle adults *(110)* Tines or spikes used for aerification can locally kill grubs *(111)* Grass species or cultivars less susceptible to damage *(112)* Specifically for Japanese beetle *(113,114)* and European chafer *(115)* Mating disruption for oriental beetle *(80)*
Billbugs	Overseeding endophytic ryegrass into certain bermudagrass *(116)* or Ky bluegrass *(117)* can reduce damage and larvae
Mole crickets (*Neoscapteriscus* spp.)	Grass species or cultivars less susceptible to damage *(112)*
Fall armyworms	Grass species or cultivars less susceptible to damage *(96,112,118,119)*
Tropical sod webworm	Grass species or cultivars less susceptible to damage *(112)*
Black cutworm	Dispose of creeping bentgrass clippings with eggs away from greens and tees *(88)* Grass species or cultivars less susceptible or resistant to damage *(112,120,121)*
Chinch bug	Grass species or cultivars less susceptible to damage *(112,122)* Southern chinch bug: Reduce mowing heights and thatch management in St. Augustine *(123)* Western chinch bug: Grass species or cultivars less susceptible to damage *(112,124,125)*

and certain plant pathogens such as *Botrytis, Pythium,* or *Fusarium* (necrotrophic pathogens) *(68)*. Constitutive or induced plant defense that effects herbivore biology or behavior would also be considered direct. Indirect defenses are those that manipulate natural enemies to assist with defenses *(67)* and only impact herbivores through the action of a natural enemy. The release of volatile organic compounds by plants for the recruitment of natural enemies was discussed previously in this chapter and in Chapter 3.

A major advantage of host plant resistance is the reduction in insecticide usage for pest management particularly in production of ornamental and turfgrass. There are also significant limitations on host plant resistance. First, plants resistant to insects or mites must have resistance to pathogens or desirable agronomic or horticultural traits. In many instances, insect resistance is a secondary evaluation for breeders of ornamentals and turfgrasses. Turfgrasses may first be selected for playability for sports turf, color, and stand qualities desired in lawns, or disease resistance. Cool season grasses such as tall fescue host nonpathogenic, fungal endophytes (*Epichlöe* or *Neotyphodium* spp.). These fungi produce alkaloids internally that convey resistance to aboveground insects *(69)* with more limited effects on below-ground herbivores like white grubs *(70,71)*. In ornamental plants, disease resistance is also important but so are esthetic characteristics. Plants are rarely exposed to just one insect or one disease during the year, and there are few if any plants that are universally resistant to all insects and diseases. Disease and insect resistance are seemingly contradictory in crapemyrtles *(72)*. Crapemyrtles resistant to the common foliar disease, powdery mildew, are typically more susceptible to crapemyrtle aphids. Turfgrasses also are commonly exposed to multiple pathogens and insects. In the context of plant defenses, the order in which the plant is attacked may influence resistance when more than one pest is involved. A review of published studies *(68)* suggests an initial plant attack by a pathogen should not affect insect performance on that plant later. However, a plant damaged by a chewing insect is predicted to significantly affect the performance of insects (sap sucking or chewing) that may later damage that plant. In general, induction of the jasmonic acid pathway by plant pathogens or insects is more likely to have a greater effect on subsequent attacks by both herbivores and pathogens than pathogens and insects that induce the salicylic acid pathway *(68)*. The complex nature of host plant resistance is why few mechanisms of resistance are well understood for pests of ornamentals and turfgrass.

Another limitation on host plant resistance is the persistence of plant traits and the speed with which they can be deployed. Insects, such as southern chinch bug, can overcome host plant resistance traits bred into St. Augustinegrass *(73)*. When Floratam St. Augustinegrass was initially released in 1973, southern chinch bugs that fed upon it could live 3 weeks or less and produced just a few eggs. By 1985, populations of southern chinch bugs were found that could damage, survive, and lay eggs on Floratam. These populations were uncommon but have since become more widespread, causing a failure of host plant resistance within 15 years of release. New cultivars that display resistance as well as some selections with tolerance are being evaluated for southern chinch bugs *(74)*. Tolerant St. Augustinegrass will grow despite heavy feeding pressure by chinch bugs and should have greater persistence than cultivars that are resistant. Evaluation and eventual deployment of varieties with insect resistance are additional limitations on host plant resistance. Resistant varieties must either come through breeding programs or be evaluations of existing varieties for resistance traits and over time in field and/or greenhouse tests. The slower response rate of host plant resistance to pests is most evident when new species invade. When an herbivore is accidentally introduced and is determined to be highly destructive, it usually takes years to evaluate plants for resistance. Adoption of host plant resistance research can also be slow. Plants or sod must be grown in nurseries or on sod farms in large numbers before they can be distributed and established. For example, ash species resistant to emerald ash borer have been identified, *(75)* but it will be years until ash species can be reestablished in urban forests. Similarly, host plant resistance data may suggest new grasses that are resistant to white grubs or mole crickets. Yet, a turfgrass renovation on a golf course or home lawn is expensive, and golf courses also lose revenue for months while the course is under renovations.

Tables 8.1 and 8.2 also outline cultural controls that exploit insect behavior or vulnerable times in the life cycle of key pests. With some exceptions, most urban landscape pests must find and secure mates once or more in a season to produce viable offspring. This is mediated through sex pheromones produced by one sex, usually the female, to recruit the other sex. A practice called mating disruption applies sex pheromones as a liquid or aerosol or as a solid pellet that releases slowly over time. The excess sex pheromone in the area interrupts the process of mate location, resulting in few females able to produce viable offspring. The pheromones must be present when adults are present and seeking mates, and applications annually or in successive years are generally needed for pest suppression. Most studies gauge success by declining or eliminating males captured in traps and by pest or damage surveys. Mating disruption has been successfully demonstrated for armored scales *(76)*, mealybugs *(77)*, wood-boring and foliage-feeding caterpillars *(78,79)*, and white grubs *(80)*. While these examples demonstrate proof of concept with important pests, mating disruption of gypsy moths *(78)* and oriental beetles *(80)* are the two successful examples relevant to urban landscapes. The attraction to plant volatiles can also be exploited in pest management. The use of trap trees for exotic borers exploits insect attraction to damaged trees or stress volatiles. Attacks on trap trees can be induced by treatment *(81,82)* or those that are naturally attacked, then left in place until the insect flight stops *(83)*. If the trap trees are destroyed, they become a biological dead end for the pest. For emerald ash borer, girdled trees used for detection can be trap trees that serve to reduce populations in the area *(81)*.

Plants in urban landscapes are subject to frequent pruning and mowing. These practices can have positive and negative effects on the abundance and biology of pests and natural enemies. Pruning can remove overwintering eggs or pupal stages of insects for pests that overwinter on limbs or in plant debris. The collection of twigs infested with twig girdlers or pruners under large trees can remove the overwintering stages of those insects *(84)*. Removal of blooms for flower-seeking species such as Japanese beetles can reduce attractiveness of those plants and damage *(85,86)*. However, not all effects of pruning have positive effects on insects. Pruning can also induce more growth, which can create opportunities for "flush feeders" and some gall-inducing species to colonize plants *(87)*. In turfgrass, mowing occurs more frequently than pruning ornamentals. Mowing, like pruning in ornamental plants, can remove egg masses of some turf-infesting caterpillars. Once eggs are removed, they can pass through the mower and clipping management becomes an important part of cultural control for black cutworms (88). Mowing height is reported as both significant and nonsignificant effects in studies in turfgrass. For example, black turfgrass ataenius larvae are more abundant when perennial ryegrass is cut at a lower (1.6 cm) height compared with a mowing height of 5.1 cm. This pattern could be driven by greater populations of predatory rove beetles and greater incidence of milky disease in the higher mown grass (89). Greater numbers of natural enemies or attacks on pests in higher mown grass are documented in other studies (90–92). The patterns are less clear for mowing height of the four common warm season turfgrasses (bermudagrass, centipedegrass, zoysiagrass, and St. Augustinegrass). Certain plant feeders (leaf hopper, spittlebugs, chinch bugs) and only certain predators (rove beetles, spiders) are more abundant in taller lawns (93).

Highlights

- Nonchemical approaches are important to the application of IPM in urban landscapes.
- Endemic natural enemies can impose moderate to high levels of mortality on pest populations.
- Conservation and augmentative biological control are ways to introduce or supplement populations of natural enemies for pest control.
- Host plant resistance is widely cited as a sustainable form of biological control, but this approach can be limited by the occurrence of multiple pests in the landscape and the costs to convert existing planting or turfgrass to new varieties.

References

1. Potter, D. A. Natural Enemies Reduce Pest Populations in Turf. *USGA Green Section Record* **1992**, *30* (6), 6–10.
2. Meineke, E. K.; Dunn, R. R.; Sexton, J. O.; Frank, S. D. Urban Warming Drives Insect Abundance on Street Trees. *PLoS One* **2013**, *8* (3), e59687. https://doi.org/10.1371/journal.pone.0059687.
3. López, R.; Potter, D. A. Ant Predation on Eggs and Larvae of the Black Cutworm and Japanese Beetle in Turfgrass. *Environmental Entomology* **2000**, *29*, 116–125.
4. Cockfield, S. D.; Potter, D. A. Predation of Sod Webworm (Lepidoptera: Pyralidae) Eggs as Affected by Chlorpyrifos Application to Kentucky Bluegrass Turf. *Journal of Economic Entomology* **1984**, *77*, 1542–1544.
5. Flint, M. L.; Dreistadt, S. H.; Clark, J. K. *Natural Enemies Handbook: The Illustrated Guide to Biological Pest Control;* University of California Press: Oakland, 1998.
6. Wheeler, A. G., Jr.; Hoebeke, E. R.; Smith, E. H. Charles Valentine Riley: Taxonomic Contributions of an Eminent Agricultural Entomologist. *American Entomologist* **2010**, *56* (1), 14–30.

7. Shrewsbury, P. M.; Leather, S. R. Using Biodiversity for Pest Suppression in Urban Landscapes. In *Biodiversity and Insect Pests: Key Issues for Sustainable Management;* Gurr, G. M., Wratten, S. D., Snyder, W. E., Eds.; John Wiley & Sons Ltd: Chichester, UK, 2012; p 350.

8. Tooker, J. F.; Hanks, L. M. Influence of Plant Community Structure on Natural Enemies of Pine Needle Scale (Homoptera: Diaspididae) in Urban Landscapes. *Environmental Entomology* **2000,** *29* (6), 1305−1311.

9. Frank, S. D.; Shrewsbury, P. M. Effect of Conservation Strips on the Abundance and Distribution of Natural Enemies and Predation of *Agrotis ipsilon* (Lepidoptera: Noctuidae) on Golf Course Fairways. *Environmental Entomology* **2004,** *33,* 1662−1672.

10. Rogers, M. E.; Potter, D. A. Potential for Sugar Sprays and Flowering Plants to Increase Parasitism of White Grubs by Tiphiid Wasps (Hymenoptera: Tiphiidae). *Environmental Entomology* **2004,** *33,* 619−626.

11. Ellis, J. A.; Walter, A. D.; Tooker, J. F.; Ginzel, M. D.; Reagel, P. F.; Lacey, E. S.; Bennett, A. B.; Grossman, E. M.; Hanks, L. M. Conservation Biological Control in Urban Landscapes: Manipulating Parasitoids of Bagworm (Lepidoptera: Psychidae) with Flowering Forbs. *Biological Control* **2005,** *34,* 99−107.

12. Rebek, E. J.; Sadof, C. S.; Hanks, L. M. Manipulating the Abundance of Natural Enemies in Ornamental Landscapes with Floral Resource Plants. *Biological Control* **2005,** *33,* 203−216.

13. Rebek, E. J.; Sadof, C. S.; Hanks, L. M. Influence of Floral Resource Plants on Control of an Armored Scale Pest by the Parasitoid *Encarsia citrina* (Craw.) (Hymenoptera: Aphelinidae). *Biological Control* **2006,** *37,* 320−328.

14. Held, D. W.; Abraham, C. M. Biology of *Larra bicolor,* a Parasitoid of Mole Crickets, and Potential Ornamental Plant as Nectar Sources. *USGA Turfgrass and Environmental Research Online* **2010,** *9* (4), 1−10.

15. Portman, S. L.; Frank, H.; McSorley, J.; Leppla, N. C. Nectar-seeking and Host-Seeking by *Larra bicolor* (Hymenoptera: Crabronidae), a Parasitoid of *Scapteriscus* Mole Crickets (Orthoptera: Gryllotalpidae). *Environmental Entomology* **2010,** *39* (3), 939−943.

16. Zemenick, A. T.; Kula, R. R.; Russo, L.; Tooker, J. F. A Network Approach Reveals Parasitoid Wasps to Be Generalized Nectar Foragers. *Arthropod-Plant Interactions* **2019,** *13,* 239−251.

17. Atlegrim, O. Exclusion of Birds from Bilberry Stands—Impact on Insect Larval Density and Damage to the Bilberry. *Oecologia* **1989,** *79,* 136−139.

18. Marquis, R. J.; Whelan, C. J. Insectivorous Birds Increase Growth of White Oak through Consumption of Leaf-Chewing Insects. *Ecology* **1994,** *75,* 2007−2014.

19. Giffard, B.; Corcket, E.; Barbaro, L.; Jactel, H. Bird Predation Enhances Tree Seedling Resistance to Insect Herbivores in Contrasting Forest Habitats. *Oecologia* **2012,** *168,* 415−424.

20. Singer, M. S.; Clark, R. E.; Lichter-Marck, I. H.; Johnson, E. R.; Mooney, K. A. Predatory Birds and Ants Partition Caterpillar Prey by Body Size and Diet Breadth. *Journal of Animal Ecology* **2017,** *86,* 1363−1371.

21. Gunnarsson, B.; Wallin, J.; Klingberg, J. Predation by Avian Insectivores on Caterpillars is Linked to Leaf Damage on Oak (Quercus *robur*). *Oecologia* **2018,** *188,* 733−741.

22. Kozlov, M. V.; Lanta, V.; Zverev, V.; Rainio, K.; Kunavin, M. A.; Zvereva, E. L. Decreased Losses of Woody Plant Foliage to Insects in Large Urban Areas Are Explained by Bird Predation. *Global Change Biology* **2017,** *23,* 4354−4364.

23. Martinson, T.; Flaspohler, D. Winter Bird Feeding and Localized Predation on Simulated Bark-Dwelling Arthropods. *Wildlife Society Bulletin* **2003,** *31* (2), 510−516.

24. Orros, M. E.; Fellowes, M. D. E. Supplementary Feeding of Wild Birds Indirectly Affects the Local Abundance of Arthropod Prey. *Basic and Applied Ecology* **2012,** *13* (3), 286–293.

25. Orros, M. E.; Thomas, R. L.; Holloway, G. J.; Fellowes, M. D. E. Supplementary Feeding of Wild Birds Affects Ground Beetle Populations in Suburban Gardens. *Urban Ecosystems* **2015,** *18,* 465–475.

26. Kaplan, I. Attracting Carnivorous Arthropods with Plant Volatiles: The Future of Biocontrol or Playing With Fire? *Biological Control* **2012,** *60* (2), 77–89.

27. Rodriguez-Saona, C.; Kaplan, I.; Braasch, J.; Chinnasamy, D.; Williams, L. Field Responses of Predaceous Arthropods to Methyl Salicylate: a Meta-Analysis and Case Study in Cranberries. *Biological Control* **2011,** *59* (2), 294–303.

28. White, J.; Johnson, D. *Vendors of Beneficial Organisms in North America;* ENTFACT-125 University of Kentucky Cooperative Extension Service, 2010; p 7.

29. Raupp, M. J.; Hardin, M. R.; Braxton, S. M.; Bull, B. B. Augmentative Releases for Aphids Control on Landscape Plants. *Journal of Arboriculture* **1994,** *20* (5), 241–249.

30. Addesso, K. M.; Witcher, A. L.; Fare, D. C. Swirski Mite Controlled-Release Sachets as a Pest Management Tool in Container Tree Production. *HortTechnology* **2018,** *28* (3), 391–398.

31. Flint, M. L.; Dreistadt, S. H. Interactions Among Convergent Lady Beetle (*Hippodamia convergens*) Releases, Aphid Populations, and Rose Cultivar. *Biological Control* **2005,** *34* (1), 38–46.

32. Dreistadt, S. H.; Flint, M. L. Melon Aphid (Homoptera: Aphididae) Control by Inundative Convergent Lady Beetle (Coleoptera: Coccinellidae) Release on chrysanthemum. *Environmental Entomology* **1996,** *25* (3), 688–697.

33. Shrewsbury, P. M.; Hardin, M. R. Evaluation of Predatory Mite (Acari: Phytoseiidae) Releases to Suppress Spruce Spider Mites, *Oligonychus ununguis* (Acari: Tetranychidae), on Juniper. *Journal of Economic Entomology* **2003,** *96* (6), 1675–1684.

34. Koppenhöfer, A. M.; Fuzy, E. M. Nematodes for White Grub Control. *USGA Turfgrass and Environmental Research Online* **2006,** *5* (19), 1–10.

35. McGraw, B. A.; Vittum, P. J.; Cowles, R. S.; Koppenhöfer, A. M. Field Evaluation of Entomopathogenic Nematodes for the Biological Control of the Annual Bluegrass Weevil, *Listronotus maculicollis* (Coleoptera: Curculionidae), in Golf Course Turfgrass. *Biocontrol Science and Technology* **2010,** *20* (2), 149–163.

36. Koppenhöfer, A. M.; Wu, S. Microbial Control of Insect Pests of Turfgrass. In *Microbial Control of Insect and Mite Pests: From Theory to Practice;* Lacey, L. A., Ed.; Academic Press: San Diego, California, 2017; pp 331–341.

37. Grewal, P. S. Factors in the Success and Failure of Microbial Control in Turfgrass. *Integrated Pest Management Reviews* **1999,** *4,* 287–294.

38. Gardner, W. A.; Noblet, R.; Schwehr, R. D. The Potential of Microbial Agents in Managing Populations of the Fall Armyworm (Lepidoptera: Noctuidae). *Florida Entomologist* **1984,** *67,* 325–332.

39. Koppenhöfer, A. M.; Klein, M. G. Microbial Control of Turfgrass Insects. In *Handbook of Turfgrass Insects;* Brandenburg, R. L., Freeman, C. P., Eds., 2nd ed.; Entomological Society of America: Maryland, 2012.

40. Prater, C. A.; Redmond, C. T.; Barney, W.; Bonning, B.; Potter, D. A. Microbial Control of the Black Cutworm (Lepidoptera: Noctuidae) in Turfgrass Using *Agrotis ipsilon* Multiple Nucleopolyhedrovirus. *Journal of Economic Entomology* **2006,** *99,* 1129–1137.

41. Bixby-Brosi, A. J.; Potter, D. A. Evaluating a Naturally Occurring Baculovirus for Extended Biological Control of the Black Cutworm (Lepidoptera: Noctuidae) in Golf Course Habitats. *Journal of Economic Entomology* **2010,** *103,* 1555–1563.

42. Krueger, S. R.; Nechols, J. R.; Ramoska, W. A. Infection of Chinch Bug, *Blissus leucopterus leucopterus* (Hemiptera: Lygaeidae), Adults from *Beauveria bassiana* (Deuteromycotina: Hyphomycetes) Conidia in Soil Under Controlled Temperature and Moisture Conditions. *Journal of Invertebrate Pathology* **1991**, *58*, 19−26.

43. Samuels, R. I.; Coracini, D. L. A.; Martins dos Santos, C. A.; Gava, C. A. T. Infection of *Blissus antillus* (Hemiptera: Lygaeidae) Eggs by the Entomopathogenic Fungi *Metarhizium anisopliae* and *Beauveria bassiana*. *Biological Control* **2002**, *23*, 269−273.

44. Thompson, S. R.; Brandenburg, R. L.; Arends, L. L. Impact of Moisture and UV Degradation on *Beauveria bassiana* (Balsamo) Vuillemin Conidial Viability in Turfgrass. *Biological Control* **2006**, *39*, 401−407.

45. Giroux, F.; Lavallée, R.; Bauce, E.; Guertin, C. Susceptibility of the Japanese Beetle, *Popillia japonica* (Newman) (Coleoptera: Scarabaeidae), to Entomopathogenic *Hypocreales* Fungi. *Phytoprotection* **2015**, *95*, 1−6.

46. Gan, H.; Churchill, A. C. L.; Wickings, K. Invisible but Consequential: Root Endophytic Fungi Have Variable Effects on below Ground Plant-Insect Interactions. *Ecosphere* **2017**, *8* (3), e01710.

47. Xia, Y.; Hertl, P.; Brandenburg, R. L. Surface and Subsurface Application of *Beauveria bassiana* for Controlling Mole Crickets (Orthoptera: Gryllotalpidae) in Golf Courses. *Journal of Agricultural and Urban Entomology* **2000**, *17*, 177−189.

48. Dingman, D. W. Geographical Distribution of Milky Disease Bacteria in the Eastern United States Based on Phylogeny. *Journal of Invertebrate Pathology* **2008**, *97*, 171−181.

49. Carson, R. *Silent Spring;* Houghton Mifflin: Boston, 1962.

50. Redmond, C. T.; Williams, D. W.; Potter, D. A. Comparison of Scarab Grub Populations and Associated Pathogens and Parasitoids in Warm- or Cool-Season Grasses Used on Transitional Zone Golf Courses. *Journal of Economic Entomology* **2012**, *105* (4), 1320−1328.

51. Redmond, C. T.; Potter, D. A. Incidence of Turf-Damaging White Grubs (Coleoptera: Scarabaeidae) and Associated Pathogens and Parasitoids on Kentucky Golf Courses. *Environmental Entomology* **2010**, *39*, 1838−1847.

52. Cappaert, D. L.; Smitley, D. R. Parasitoids and Pathogens of Japanese Beetle (Coleoptera: Scarabaeidae) in Southern Michigan. *Environmental Entomology* **2002**, *31*, 573−580.

53. Redmond, C. T.; Potter, D. A. Lack of Efficacy of In Vivo- and Putatively In Vitro-produced *Bacillus popilliae* against Field Populations of Japanese Beetle (Coleoptera: Scarabaeidae) Grubs in Kentucky. *Journal of Economic Entomology* **1995**, *88*, 846−854.

54. Koppenhöfer, A. M.; Wilson, M.; Brown, I.; Kaya, H.; Gaugler, R. Biological Control Agents for White Grubs (Coleoptera: Scarabaeidae) in Anticipation of the Establishment of the Japanese Beetle in California. *Journal of Economic Entomology* **2000**, *93*, 71−80.

55. Tofangsazi, N.; Cherry, R. H.; Arthurs, S. P. Efficacy of Commercial Formulations of Entomopathogenic Nematodes against Tropical Sod Webworm, *Herpetogramma phaeopteralis* (Lepidoptera: Crambidae). *Journal of Applied Entomology* **2014**, *138*, 656−661.

56. Koppenhöfer, A. M.; Fuzy, E. M. Long-term Effects and Persistence of *Steinernema Scarabaei* Applied for Suppression of *Anomala orientalis* (Coleoptera: Scarabaeidae). *Biological Control* **2009**, *48*, 63−72.

57. Barbara, K. A.; Buss, E. A. Augmentative Applications of *Steinernema Scapterisci* (Nematoda: Steinernematidae) for Mole Cricket (Orthoptera: Gryllotalpidae) Control on Golf Courses. *Florida Entomologist* **2006**, *89*, 257−262.

58. Barbara, K. A.; Buss, E. A. Integration of Insect Parasitic Nematodes (Rhabditida: Steinernematidae) with Insecticides for Control of Pest Mole Crickets (Orthoptera: Gryllotalpidae: *Scapteriscus* spp.). *Journal of Economic Entomology* **2005**, *98*, 689−693.

59. Koppenhöfer, A. M.; Grewal, P. S.; Kaya, H. K. Synergism of Imidacloprid and Entomopathogenic Nematodes against White Grubs: the Mechanism. *Entomologia Experimentalis et Applicata* **2000,** *94,* 283−293.
60. Dobbs, E. K.; Potter, D. A. Naturalized Habitat on Golf Courses: Source or Sink for Natural Enemies and Conservation Biological Control? *Urban Ecosystems* **2016,** *19,* 899−914.
61. Freeman Long, R.; Corbet, A.; Lamb, C.; Reberg-Horton, C.; Chandler, J.; Stimmann, M. Beneficial Insects Move from Flowering Crops to Nearby Crops. *California Agriculture* **1998,** *52* (5), 23−26.
62. Landis, D. A.; Wratten, S. D.; Gurr, G. M. Habitat Management to Conserve Natural Enemies of Arthropod Pests in Agriculture. *Annual Review of Entomology* **2000,** *45* (1), 175−201.
63. Sourakov, A. Trophic Interactions Involving *Herpetogramma phaeopteralis* (Lepidoptera: Pyralidae) and *Passiflora incarnata* (Passifloraceae). *Florida Entomologist* **2008,** *91* (1), 136−138.
64. Helms, K. R.; Vinson, S. B. Apparent Facilitation of an Invasive Mealybug by an Invasive Ant. *Insectes Sociaux* **2003,** *50,* 403−404.
65. Vanek, S. J.; Potter, D. A. Ant-exclusion to Promote Biological Control of Soft Scales (Hemiptera: Coccidae) on Woody Landscape Plants. *Environmental Entomology* **2010,** *39* (6), 1829−1837.
66. Painter, R. H. *Insect Resistance in Crop Plants;* University of Kansas Press: Lawrence, 1951.
67. Stout, M. J. Reevaluating the Conceptual Framework for Applied Resaearch on Host-Plant Resistance. *Insect Science* **2013,** *20,* 263−272.
68. Moreira, X.; Abdala-Roberts, L.; Castagneyrol, B. Interactions between Plant Defence Signalling Pathways: Evidence from Bioassays With Insect Herbivores and Plant Pathogens. *Journal of Ecology* **2018,** *106,* 2353−2364.
69. Richmond, D. S.; Shetlar, D. J. Hairy Chinch Bug (Hemiptera: Lygaeidae) Damage, Population Density, and Movement in Relation to the Incidence of Perennial Ryegrass Infected by *Neotyphodium* Endophytes. *Journal of Economic Entomology* **2000,** *93* (4), 1167−1172.
70. Potter, D. A.; Patterson, C. G.; Redmond, C. T. Influence of Turfgrass Species and Tall Fescue Endophyte on Feeding Ecology of Japanese Beetle and Southern Masked Chafer Grubs (Coleoptera: Scarabaeidae). *Journal of Economic Entomology* **1992,** *85,* 900−909.
71. Koppenhöfer, A. M.; Cowles, R. S.; Fuzy, E. M. Effects of Turfgrass Endophytes (Clavicipitaceae: Ascomycetes) on White Grub (Coleoptera: Scarabaeidae) Larval Development and Field Populations. *Environmental Entomology* **2003,** *32* (4), 895−906.
72. Mizell, R. F.; Knox, G. W. Susceptibility of Crape Myrtle, *Lagerstroemia inica* L., to the Crapemyrtle Aphid (Homoptera: Aphididae) in North Florida. *Journal of Entomological Science* **1993,** *28* (1), 1−7.
73. Busey, P.; Center, B. J. Southern Chinch Bug (Hemiptera: Heteroptera: Lygaeidae) Overcomes Resistance in St. Augustinegrass. *Journal of Economic Entomology* **1987,** *80* (3), 608−611.
74. Milla-Lewis, S. R.; Youngs, K. M.; Arrellano, C.; Cardoza, Y. J. Tolerance in St. Augustinegrass Germplasm against *Blissus insularis* Barber (Hemiptera: Blissidae). *Crop Science* **2017,** *57,* S26−S36.
75. Tanis, S. R.; Mccullough, D. G. Host Resistance of Five *Fraxinus* Species to *Agrilus planipennis* (Coleoptera: Buprestidae) and Effects of Paclobutrazol and Fertilization. *Environmental Entomology* **2015,** *44* (2), 287−299.
76. Vacas, S.; Alfaro, C.; Navarro-Llopis, V.; Primo, J. Mating Disruption of California Red Scale, *Aonidiella aurantii* Maskell (Homoptera: Diaspididae), Using Biodegradable Mesoporous Pheromone Dispensers. *Pest Management Science* **2010,** *66,* 745−751.

77. Sharon, R.; Zahavi, T.; Sokolsky, T.; Sofer-Arad, C.; Tomer, M.; Kedoshim, R.; Harari, A. R. Mating Disruption Method against the Vine Mealybug, *Planococcusficus*: Effect of Sequential Treatment on Infested Vines. *Entomologia Experimentalis et Applicata* **2016,** *161,* 65−69.

78. Leonhardt, B. A.; Mastro, V. C.; Leonard, D. S.; McLane, W.; Reardon, R. C.; Thorpe, K. W. Control of Low-Density Gypsy Moth (Lepidoptera: Lymantriidae) Populations by Mating Disruption with Pheromone. *Journal of Chemical Ecology* **1996,** *22,* 1255−1272.

79. Hegazi, E. M.; Khafagi, W. E.; Konstantopoulou, M. A.; Schlyter, F.; Raptopoulos, D.; Shweil, S.; Abd El-Rahman, S.; Atwa, A.; Ali, S. E.; Tawfik, H. Suppression of Leopard Moth (Lepidoptera: Cossidae) Populations in Olive Trees in Egypt through Mating Disruption. *Journal of Economic Entomology* **2010,** *103* (5), 1621−1627.

80. Koppenhöfer, A. M.; Polavarapu, S.; Fuzy, E. M.; Zhang, A.; Ketner, K.; Larsen, T. Mating Disruption of Oriental Beetle (Coleoptera: Scarabaeidae) in Turfgrass Using Microencapsulated Formulations of Sex Pheromone Components. *Environmental Entomology* **2005,** *34* (6), 1408−1417.

81. Mercader, R. J.; McCullough, D. G.; Storer, A. J.; Bedford, J. M.; Poland, T. M.; Katovich, S. Evaluations of the Potential Use of a Systemic Insecticide and Girdled Trees in Area Wide Management of the Emerald Ash Borer. *Forest Ecology and Management* **2015,** *350,* 70−80.

82. Reding, M. E.; Ranger, C. M.; Oliver, J. B.; Schultz, P. B.; Youssef, N. N.; Bray, A. M. Ethanol-injection Induces Attacks by Ambrosia Beetles (Coleoptera: Curculionidae: Scolytinae) on a Variety of Tree Species. *Agriculture and Forest Entomology* **2017,** *19,* 34−41.

83. Mizell, R.; Riddle, T. C. Evaluation of Insecticides to Control the Asian Ambrosia Beetle, *Xylosandrus crassiusculus. Proceedings of the Southern Nursery Association Research Conference* **2004,** *49,* 152−155.

84. Johnson, W. T.; Lyon, H. H. *Insects that Feed on Trees and Shrubs,* 2nd ed.; Cornell University Press: Ithaca, 1991.

85. Vitullo, J. M.; Sadof, C. S. Effects of Pesticide Applications and Cultural Controls on Efficacy of Control for Adult Japanese Beetles (Coleoptera: Scarabaeidae) on Roses. *Journal of Economic Entomology* **2007,** *100* (1), 95−102.

86. Held, D. W.; Potter, D. A. Floral Affinity and Benefits of Dietary Mixing with Flowers for a Polyphagous Scarab, Popillia Japonica Newman. *Oecologia* **2004,** *140,* 312−320.

87. Held, D. W.; Wheeler, C.; Boyd, D. W., Jr. Kaolin Particle Film Prevent Galling by *Gynaikothrips uzeli. Plant Health Progress* **2009;** https://doi.org/10.1094/PHP-2009-0407-02-RS.

88. Williamson, R. C.; Potter, D. A. Oviposition of Black Cutworm (Lepidoptera: Noctuidae) on Creeping Bentgrass Putting Greens and Removal of Eggs by Mowing. *Journal of Economic Entomology* **1997,** *90,* 590−594.

89. Rothwell, N. L.; Smitley, D. R. Impact of Golf Course Mowing Practices on *Ataenius spretulus* (Coleoptera: Scarabaeidae) and its Natural Enemies. *Environmental Entomology* **1999,** *28* (3), 358−366.

90. Potter, D. A.; Powell, A. J.; Spicer, P. G.; Williams, D. W. Cultural Practices Affect Root-Feeding White Grubs (Coleoptera: Scarabaeidae) in Turfgrass. *Journal of Economic Entomology* **1996,** *89,* 156−164.

91. Hong, S. C.; Held, D. W.; Williamson, R. C. Beneficial Arthropods and Predation on Black Cutworm Larvae (*Agrotis ipsilon*) in Close Mown Creeping Bentgrass. *Florida Entomologist* **2011,** *94* (3), 714−715.

92. Dobbs, E. K.; Potter, D. A. Conservation Biological Control and Pest Performance in Lawn Turf: Does Mowing Height Matter? *Environmental Management* **2014,** *53,* 648–659.

93. Joseph, S. V.; Braman, S. K. Influence of Plant Parameters on Occurrence and Abundance of Arthropods in Residential Turfgrass. *Journal of Economic Entomology* **2009,** *102,* 1116–1122.

94. Lemke, H. D.; Raupp, M. J.; Shrewsbury, P. M. Efficacy and Costs Associated with the Manual Removal of Bagworms, *Thyridopteryx ephemeraeformis,* from Leyland Cypress. *Journal of Environmental Horticulture* **2005,** *23* (3), 123–126.

95. Klingeman, W. E. Bagworm Survival and Feeding Preferences as Indicators of Resistance Among Maples. *Journal of Environmental Horticulture* **2002,** *20* (3), 138–142.

96. Braman, S. K.; Carr, E. R.; Quick, J. C. *Canna* Spp. Cultivar Response to the Lesser Canna Leafroller, *Geshna cannalis,* (Quaintance), and the Japanese Beetle, *Popillia japonica* (Newman). *Journal of Environmental Horticulture* **2011,** *29* (2), 87–90.

97. Reinert, J. A.; George, S. W.; Mackay, W. A.; Davis, T. D. Resistance Among Lantana Cultivars to the Lantana Lace Bug, *Teleonemia scrupulosa* Hemiptera: Tingidae). *Florida Entomologist* **2006,** *89* (4), 449–454.

98. Held, D. W. Relative Susceptibility of Woody Landscape Plants to Japanese Beetle (Coleoptera: Scarabaeidae). *Journal of Arboriculture* **2004,** *30* (6), 328–335.

99. Condra, J. M.; Brady, C. M.; Potter, D. A. Resistance of Landscape-Suitable Elms to Japanese Beetle, Gall Aphids, and Leaf Miners, with Notes Life History of *Orchestes alni* and *Agromyza aristata* in Kentucky. *Arboriculture and Urban Forestry* **2010,** *36* (3), 101–109.

100. Kreuger, B.; Potter, D. A. Diel Feeding Activity and Thermoregulation by Japanese Beetles (Coleoptera: Scarabaeidae) within Host Plant Canopies. *Environmental Entomology* **2001,** *30,* 172–180.

101. Switzer, P. V.; Cumming, R. M. Effectiveness of Hand Removal for Small-Scale Management of Japanese Beetles (Coleoptera: Scarabaeidae). *Journal of Economic Entomology* **2014,** *107* (1), 293–298.

102. Nair, S.; Braman, S. K. A Scientific Review on the Ecology and Management of the Azalea Lace Bug *Stephanitis pyroides* (Scott) (Tingidae: Hemiptera). *Journal of Entomological Science* **2012,** *47* (3), 247–263.

103. Guam, W. G.; Giliomee, J. H.; Pringle, J. H. Resistance of Some Rose Cultivars to the Western Flower Thrips, *Frankliniella occidentalis* (Thysanoptera: Thripidae). *Bulletin of Entomological Research* **1994,** *84,* 487–492.

104. Bergh, C. J.; Le Blanc, J. P. R. Performance of Western Flower Thrips (Thysanoptera: Thripidae) on Cultivars of Miniature Rose. *Journal of Economic Entomology* **1997,** *90* (2), 679–688.

105. Schuch, U. K.; Redak, R. A.; Bethke, J. A. Cultivar, Fertilizer, and Irrigation Affect Vegetative Growth and Susceptibility of Chrysanthemum to Western Flowers Thrips. *Journal of the American Society of Horticultural Science* **1998,** *123* (4), 727–733.

106. Herrin, B.; Warnock, D. Resistance of Impatiens Germplasm to Western Flower Thrips Feeding Damage. *HortScience* **2002,** *37* (5), 802–804.

107. Held, D. W.; Wheeler, C.; McLaurin, W. Cultural Practices for Removal of Wax Scales and Sooty Mold from Ornamentals. *Proceedings of the Southern Nursery Association Research Conference* **2006,** *51,* 141–144.

108. Stoyenoff, J. L. Plant Washing as a Pest Management Technique for Control of Aphids (Homoptera: Aphididae). *Journal of Economic Entomology* **2001,** *94* (6), 1492–1499.

109. Braman, S. K.; Buntin, G. D.; Oetting, R. D. Species and Cultivar Influences on Infestation by and Parasitism of a Columbine Leafminer (*Phytomyza aquilegivora* Spencer). *Journal of Environmental Horticulture* **2005,** *23* (1), 9–13.

110. Billeisen, T. L.; Brandenburg, R. L. Biology and Management of the Sugarcane Beetle (Coleoptera: Scarabaeidae) in Turfgrass. *Journal of Integrated Pest Management* **2014,** *5* (4). https://doi.org/10.1603/IPM14008.

111. Cranshaw, W. S.; Zimmerman, R. J. Biological, Mechanical, and Chemical Control of Turfgrass-Infesting Scarabs in Colorado. *Southwestern Entomologist* **1989,** *14* (4), 351–355.

112. Reinert, J. A.; Engelke, M. C.; Read, J. C. Host Resistance to Insects and Mites, a Review- A Major IPM Strategy in Turfgrass Culture. *Acta Horticulturae* **2004,** *661,* 463–486.

113. Braman, S. K.; Raymer, P. L. Impact of Japanese Beetle (Coleoptera: Scarabaeidae) Feeding on Seashore Paspalum. *Journal of Economic Entomology* **2006,** *99* (5), 1699–1704.

114. Wood, T. N.; Richardson, M.; Potter, D. A.; Johnson, D. T.; Wiedenmann, R. N.; Steinkraus, D. C. Ovipositional Preferences of the Japanese Beetle (Coleoptera: Scarabaeidae) Among Warm- and Cool-Season Turfgrass Species. *Journal of Economic Entomology* **2009,** *102* (6), 2192–2197.

115. Bughrara, S. S.; Smitley, D. R.; Cappaert, D. European Chafer Grub Feeding on Warm-Season and Cool-Season Turfgrasses, Native Praire Grasses, and Pennsylvania Sedge. *Horttechnology* **2008,** *18* (3), 329–333.

116. Huang, T. I.; Buss, E. A. *Sphenophorus venatus Vestitus* (Coleoptera: Curculionidae) Preference for Bermudagrass Cultivars and Endophytic Perennial Ryegrass Overseed. *Florida Entomologist* **2013,** *96* (4), 1628–1630.

117. Richmond, D. S.; Niemczyk, H.,D.; Shetlar, D. J. Overseeding Endophytic Perennial Ryegrass into Stands of Kentucky Bluegrass to Manage Bluegrass Billbug (Coleoptera: Curculionidae). *Journal of Economic Entomology* **2000,** *93* (6), 1662–1668.

118. Ball, O. J. P.; Coudron, T. A.; Tapper, B. A.; Davies, E.; Trently, D.; Bush, L. P.; Gwinn, K. D.; Popay, A. J. Importance of Host Plant Species, *Neotyphodium* Endophyte Isolate, and Alkaloids on Feeding by *Spodoptera frugiperda* (Lepidoptera: Noctuidae) Larvae. *Journal of Economic Entomology* **2006,** *99* (4), 1462–1473.

119. Reinert, J. A.; Engelke, M. C. Resistance in Zoysiagrass (*Zoysia* spp.) to the Fall Armyworm (*Spodoptera frugiperda*) (Lepidoptera: Noctuidae). *Florida Entomologist* **2010,** *93* (2), 254–259.

120. Hong, S. C.; Obear, G. R.; Liesch, P. J.; Held, D. W.; Williamson, R. C. Suitability of Creeping Bentgrass and Bermudagrass Cultivars for Black Cutworms and Fall Armyworms (Lepidoptera: Noctuidae). *Journal of Economic Entomology* **2015,** *108* (4), 1954–1960.

121. Hong, S. C.; Williamson, R. C. Suitability of Various Turfgrass Species and Cultivars for Development and Survival of Black Cutworm (Lepidoptera: Noctuidae). *Journal of Economic Entomology* **2006,** *99* (3), 850–857.

122. Rangasamy, M.; McAuslane, H. J.; Cherry, R. H.; Nagata, R. T. Categories of Resistance in St. Augustinegrass Lines to Southern Chinch Bug (Hemiptera: Blissidae). *Journal of Economic Entomology* **2006,** *99* (4), 1446–1451.

123. Kaur, N.; Gillett-Kaufman, J. L.; Gezan, S. A.; Buss, E. A. Association between *Blissus insularis* Densities and St Augustinegrass Lawn Parameters in Florida. *Crop Forage Turfgrass Manage* **2016,** *2.* https://doi.org/10.2134/cftm2016.0015.

124. Gulsen, O.; Heng-Moss, T.; Shearman, R.; Baenziger, P. S.; Lee, D.; Baxendale, F. P. Buffalograss Germplasm Resistance to *Blissus occiduus* (Hemiptera: Lygaeidae). *Journal of Economic Entomology* **2004,** *97* (6), 2101–2105.

125. Anderson, W. G.; Heng-Moss, T. M.; Baxendale, F. P. Evaluation of Cool- and Warm-Season Grasses for Resistance to Multiple Chinch Bug (Hemiptera: Blissidae) Species. *Journal of Economic Entomology* **2006,** *99* (1), 203–211.

Chapter 9

Insecticides: A balance of plant protection and environmental stewardship

Introduction

About 85%–95% of American households have at least one pesticide (*1,2*). For perspective, this is similar to the percent of households with computer and Internet use in 2015 (*3*). Pesticides are chemicals designed to kill unwanted plants, plant diseases, insects, and mites. Under the umbrella of pesticides, insecticides and miticides are those designed to specifically kill arthropods. In the United States, about 61% of households have insecticides, which is more than twice the number of households with herbicides (*1*). Insecticides are designed to make short-term reductions in populations of pest insects and mites. However, science has not quite perfected the ability to selectively kill individuals of one species living among dozens of other innocuous or beneficial species. Therefore, insecticide applications have consequences for pests, and nontarget animals and plants in the community. This chapter will explore the basics of insecticides and their evolution, and the unintended consequences reported from insecticide use. Since this chapter will focus on issues and concerns with the pesticide use in urban landscapes, there will be limited reference to pesticide names. A listing of specific insecticides for pests would render sections of this chapter obsolete as products change. There are excellent state and regional resources regularly updated and available online that better serve this purpose. One example, the southeastern pest control guide (*4*), is a resource for insecticides and miticide registered for use in the southern United States. Although focused on southern states, it may be a useful starting point if a similar resource is not available in your state or region.

Evolution of insecticides

Some terms are needed to understand the issues. First, every insecticide product applied in urban landscapes is a formulation of the active ingredient with some liquid or solid carrier. The active ingredient is the most biologically active component. This idea is familiar because it has similarity to over-the-

Urban Landscape Entomology. https://doi.org/10.1016/B978-0-12-813071-1.00009-9
185

counter medicines. When purchasing something for a headache, you may recognize a brand name but the same medicine (active ingredient) is likely available in a range of brand name and generic products. Like medicines, the active ingredient is only one component of that product, with the remaining components included to make it easier to digest or even taste better (Box 9.1).

BOX 9.1 What is IRAC?
- Insecticide Resistance Action Committee
- A team of international scientists that categorizes insecticides and miticide using a system of numbers and letters (5)
- Like drugs, active ingredients can be classified chemically (chemical structure) and physiologically (how they act)
- IRAC use numbers to represent modes of action and letters to indicate chemical classification
- Information available online or through their app

With insecticides, these are called inactive ingredients, but they too may be needed to get a behavioral or physiological response from the insects. For example, active ingredients can be formulated on food and applied so that the insects inadvertently consume the active ingredient while feeding. In some formulations, the active ingredient would not work without the help of inactive ingredients. Baits represent one common route of exposure, ingestion or oral, for how insects are exposed. The other routes are inhalation and contact. Ingestion, contact, and inhalation represent how an insect is exposed and how the insecticide enters the insect, but it does not represent where they act. If our hypothetical bait contains the active ingredient abamectin, it would target the nerves and muscles once inside the insect (5). These target sites for activity are called the modes of action. There are currently more than 30 modes of action for insecticides and miticides. In the 1990s, the Food Quality Protection Act was a catalyst to drive the development of new products representing novel modes of action, reduced mammalian toxicity, and lower use rates. Most insecticides on the shelf at the wholesaler or your local box store or garden center are not the same as those used 30-40 years ago. For example, in the 1970s and 1980s, turfgrass pests were controlled with mainly organophosphate and carbamate insecticides. Those two classes of insecticides were applied at 3.4—9 kg of active ingredient per hectare (3—8 lbs of active ingredient per acre). Newer active ingredients for white grubs are applied at 0.03—2.2 kg of active ingredient per hectare (0.026—1.96 lbs of active ingredient per acre) (6). Furthermore, many of those older products were classified as moderately to very highly toxic by the United States Environmental Protection Agency (EPA). Most of the newer materials are classified as slightly to practically nontoxic on the same scale. This supports the earlier assertion that insecticides used today in urban landscapes are being applied at lower doses, with reduced toxic impacts, yet are highly effective for pest control (6). As we will see later, this does not mean that they are all environmental benign, but this evolution was a step in the right direction.

Approaches to using insecticides

Insecticides are thoroughly labeled to provide the user with guidance and regulatory information on the proper and legal use of the products. If insecticides are needed for a pest, it is not a time to be an independent thinker or to be experimenting. The label is truly the law and must be followed. There are patterns that emerge for the common groups of pests discussed in Chapters 6 and 7 (Table 9.1). Insecticides applied to foliage of grasses or ornamental plants may have either contact or systemic activity. Contact is somewhat self-explanatory; the insect or mite must come in contact with the fresh spray or toxic residue. A residue is what persists after an insecticide is applied and toxicity typically declines against pests with increasing time after the application. The route of exposure (contact or ingestion) does not indicate the mode of action. For example, the synthetic pyrethroids have contact activity and a mode of action in the nervous system. Insecticide soaps, oils, and certain plant-based insecticides also mainly have contact activity but have modes of action such as growth regulators and feeding deterrents. Environmental conditions

TABLE 9.1 Approaches to managing landscape insects with insecticides.

Target insects	Management approach
White grubs, mole crickets, and other soil insects	Soil insecticides applied before egg hatch (preventive) or after egg hatch (curative) Systemic insecticides Sprays to kill emerging or egg-laying adults
Foliage-feeding insects and mites	Insecticides\Miticides applied to foliage Systemic insecticides Oil sprays targeting egg masses
Scale insects	Oil sprays targeting overwintering females Systemic insecticides Foliar sprays at crawler hatch
Borers	Bark/truck sprays 2 weeks after first trap capture (7) Systemic insecticides
Leafminers	Sprays to target the egg-laying females Systemic insecticides
Gall inducers	Canopy sprays to target adults Systemic insecticides

(temperature, UV light, rainfall) and the chemical properties of the insecticides are important in determining residue persistence and effectiveness. Penetration into the plant or insect cuticle can also represent barriers to effectiveness for contact insecticides. Research with scale insects suggests their waxy coverings may not just be a generic barrier but a selective barrier to contact insecticides. Settled crawlers of armored scales are more susceptible to horticultural oils and settled soft scales are more susceptible to insecticidal soap (8). These differences are another reason why correct identification of scale insects to family is important for management at the crawler stage.

Systemic and translaminar insecticides penetrate into the leaves (foliar applied), or are taken up by the roots and distributed throughout the plant. Translaminar properties of insecticides indicate movement with the leaf but not movement between leaves. They are applied to the top of leaves, then penetrate the cuticle. Penetration of systemic materials into and through the leaf cuticle can be a major limitation on translaminar movement of insecticides (9,10). Interactions with the leaf cuticle also determine if a contact or systemic (translaminar) insecticide will be absorbed or only surface active (9,10). This is not just a waxy versus nonwaxy leaf issue. Buckholz (9) showed that leaves with similar foliar wax deposits range in leaf permeability by factors of 40–4000. Leaf permeability is one reason surfactants are added when applying insecticides to certain ornamental plants. Although leaf waxiness is often cited as the need for a leaf penetrant waxiness of leaves is clearly not the best determinant of when it should be added into solution with an insecticide. Once inside the cells, systemic and translaminar insecticides get distributed inside or between cells (10) and can kill insects that feed on the underside of the leaves. Once inside the leaves there are different internal locations that serve as end points where active ingredients reside or possibly accumulate. Insecticides can flow through the spaces between plant cells (apoplast), pass through the cells, and stored within cells. Insects like aphids or whiteflies that move their long stylets between cells would be exposed to materials that are present in those spaces. However, mites with short mouthparts consume cell contents and would have greater exposure to materials that accumulate within cells. Understanding where the insecticide accumulates or moves within plant tissues can predict activity against different groups of pests. Systemic activity of insecticides is also correlated with greater persistence in treated plant tissues. One application of some systemic insecticides to turfgrass or landscape trees and shrubs can persist and have efficacy against pests for months, if not for 1–2 years after application (11–15).

Systemic insecticides are now widely used for most urban landscape pests. Insecticide chemists can use various chemical characteristics (low log P or octanol/water partitioning) to predict if an active ingredient will be systemic in plants and in which tissues it may move (10,16). The neonicotinoid class of insecticides (16,17) and the anthranilic diamides (18) are two common insecticide classes used in urban landscapes with systemic activity. These products are applied to the root zone, by trunk injection, or to plant foliage according to the label, then are moved throughout the plant. Newer insecticides have been marketed for extended residual control. Two anthranilic diamides, chlorantraniliprole and cyantraniliprole, are noted for having longer residuals in turfgrass relative to synthetic pyrethroids. One study (15) noted ≥87% mortality of black cutworm larvae on creeping bentgrass treated with either chlorantraniliprole or thiamethoxam, a neonicotinoid, 88 days before the larvae were exposed. Another study reported ≥84% mortality of tropical sod webworms on a St. Augustine grass lawn treated 5 weeks prior to exposure with chlorantraniliprole (19). The chemical properties of the neonicotinoid imidacloprid predict that it would circulate in the plant xylem and easily penetrate plant membranes (20). Even within a chemical class, active ingredients can vary in their uptake usually based difference on their water solubility (16). For example, dinotefuran, a highly soluble neonicotinoid insecticide, is effective against many armored scales when applied to soil (21). In some instances, placement of insecticides directly inside trees can overcome barriers in plants for some insecticides. For example, emamectin benzoate is not very water soluble and immobile in soils, yet it has systemic and translaminar activity. This active ingredient cannot be as successfully delivered by foliar or soil application as it can when applied by trunk injection (22). Trunk injection systems (22) range from short duration injections to slow drip systems analogous to humans being given IV fluids (Fig. 9.1). Trunk injections are used in suburban and urban areas where spraying large urban trees is often difficult and could create unwanted spray drift, or where concern about groundwater contamination restricts soil applications. Some arborists outright reject trunk injection because it requires wounding trees during treatment. However, wound recovery is a normal part of plant physiology and healthy trees properly trunk injected can successfully compartmentalize those wounds with no increase in disease or rot (22,23). Soil and tree injections have comparable efficacy against pests when the same active ingredients are applied to the same tree species (14). However, trunk injections may speed the onset of lethal concentrations in foliage for insecticides that move slowly from the soil into the plant. Soil applied imidacloprid took 8–12 weeks to reach lethal levels in foliage of three tree species compared to 1–4 weeks when imidacloprid was trunk injected (25).

Another concern with trunk and soil injections is rate determination for small versus large trees. Insecticide labels will usually indicate the amount of active ingredient per centimeter (inch) of trunk diameter at breast height (DBH). Research in hemlock has determined the optimal rate by tree size class to deliver a lethal dose within the canopy to control hemlock woolly adelgids (25). This study (25) determined that current recommendations for the smaller and large tree sizes are likely overdosing trees with insecticide using the DBH calculation. Several studies have determined the amount and location of

FIGURE 9.1 A tree in China receiving a specialized tree injection analogous to fluids delivered by an IV in animals.

systemic insecticide in treated trees (*26–28*), but this is one of the only studies to carefully calibrate recommendations to foliar concentrations of insecticide and efficacy. Many systemic products used in urban landscapes, however have similar labeling and may also be overdosing trees in certain size categories. This is a concern because the use rates of systemic insecticides for turfgrass and ornamental plants are generally greater than on crop plants (*29,30*) and overdosing could create greater hazards to beneficial insect species.

Concerns with insecticide use

The concerns with insecticide use fall into the categories of resistance, environmental contamination, replacement, resurgence, and impacts on beneficial insects and mites. *Resistance* is the condition where some individuals in a population are able to survive an insecticide application that is lethal to most individuals in the population. It is not exclusive to insecticides, but can also occur in response to any pest management tactic that places selection pressure on populations. Resistance is also not simply product failure, which can result from a number of different factors including human error. Human error is most commonly seen in either using the incorrect product, an incorrect timing, or incorrect application rate. A resistant insect population cannot be separated based on appearance because the most common mechanisms for insecticide resistance occur inside the insect. Some basic principles of inheritance are needed to understand the development of insecticide-resistant populations. Insect populations vary in their genetics between populations such that some populations are even said to contain cryptic or biotype species (31–33). Different insect populations have unique attributes, usually related to insecticide resistance such as the Q biotype of whiteflies (32). These unique biotypes can interbreed with "normal" (nonresistant) biotypes. Insecticide resistant is not absolute, but best described as a proportion or a percentage. An insecticide susceptible population will typically have a low percentage of individuals with a unique biotype or behavior. When the same insecticides are either frequently used or persist longer in the environment, a unique biotype for insecticide resistance can survive while others perish. The characteristic that makes these insects resistant is passed from mother to offspring (heritable) giving the next generation a potential advantage. Over multiple generations, this resistant biotype becomes a greater percentage of the population. If one control method is used, the population transitions to a condition where the individuals that were previously in the minority are now the majority. The control method does not have to be an insecticide as insect populations have developed resistance to other pest management tactics (e.g., pheromones, insect growth regulators). This process as presented here is simplified but it outlines how a resistant population develops over time.

Insecticide resistance is only quantifiable relative to other populations of the same species. Since resistance is relative, it is not possible for someone to walk onto a property and proclaim that the population of insects or mites is resistant. Laboratory or greenhouse experiments compare doses of an insecticide needed to kill the suspected resistant population(s) to a population with limited or not insecticide exposure using a resistance ratio. Resistance ratios enable researchers to determine the extent (2×, 10×, or 500×) of resistance. Resistance is most likely to develop where the same population of insects is exposed repeatedly to the same insecticide mode of action. Removing that particular selection pressure or diversify the tools used for management can allow susceptible individuals to return, mate with resistant individuals. If integrated pest management is applied over time, a population may regain susceptibility to a particular mode of action. Prevention of resistance development was a catalyst for the creation of IRAC and the IRAC classification on insecticide labels. The goal of resistance management is to use different modes of action against pests with multiple generations per year and high numbers of offspring, which are prone to develop resistance. The incidence of resistance is also not just a problem for academics and scientists at chemical companies, it can affect the bottom line of anyone doing pest management. Borel (34) provides data showing that increased incidence of insecticide resistance overlays well with increased costs to research, develop, and register new active ingredients. For example, in 1995 it cost about $150 million (USD) to bring a product to market. Those costs increased by more than $100 million (USD) in 20 years coincident with an increase of insect species developing resistance. It is not feasible or realistic to think that companies will continue to innovate and market new active ingredients for end users, if those end users are not being good stewards of those products.

Several pests in urban landscapes are reported to have resistant populations with the most common examples being pests of turfgrass (chinch bugs, annual bluegrass weevil, fall armyworms) (35). Host plants and associated microbes play a role in insecticide resistance. First, some internal mechanisms for insecticide resistance are inherent in plant-feeding insects to enable them to detoxify plant defenses. Insects have enzymes that assist in degrading or eliminating toxins from their body. In Japanese beetles, for example, feeding can induce multiple groups of enzymes (36). This is analogous to alcohol dehydrogenase in humans. Without that enzyme, drinking your favorite adult beverage could be lethal. Many of the same groups of enzymes can be activated when insects are exposed to insecticides. When present and activated in insects, these detoxification enzymes work to degrade or eliminate insecticides in the same way as degrading plant toxins. These enzymes are common in insects (Japanese beetles, mole crickets, chinch bugs, caterpillars) and also occur in natural enemies. Second, new research suggests that microbial endophytes inside plants may pass to insects and convey resistance. Southern chinch bugs are one of the most well-studied examples of this phenomenon. *Burkholderia* bacteria can live systemically inside St. Augustine grass. As they feed, southern chinch bugs acquire the bacteria and they are stored in special pouches in the gut (37). Individual southern chinch bugs treated with an antibiotic are threefold more susceptible to bifenthrin than nontreated ones. Those treated with antibiotics also have less *Burkholderia* in their guts (38). This suggests that St. Augustine grass, when colonized by the bacteria *Burkholderia*, likely contributes to the onset of insecticide resistance in some populations.

Environmental contamination occurs when an insecticide application directly or indirectly is detected at harmful levels in a natural resource, usually soil or water. Harmful levels of contaminants in soil and water set by the federal government are called tolerances. However, detection at any level is often perceived as harmful or threatening by the general public. Insecticide persistence in soil is measured by half life, or number of days required to degrade half of the material. Most of the common insecticides have half lives of 1–3 months in soil. However, the historically longer residual insecticides like DDT and its metabolites have half lives of 5–20 years. This likely explains why DDT and its metabolites are still detected in soils in certain urban landscapes more than 40 years after uses were banned (39,40). Some have claimed that greater insect activity or diversity in urban landscapes and golf courses is related to the final degradation of these insecticides in urban soils (*41*). Urban soils are not sampled as commonly for contaminants as surface water. The United States Geological Survey samples urban streams across the country and reports the detection levels in water and stream sediments (*42*). Surveys consistently show insecticides used in urban areas are being transported into urban streams. These detections may result from the incidental transport following large rain events or when insecticides are misapplied to impervious surfaces or overdosed. Two percent or less of imidacloprid granular and spray formulations applied to bermudagrass is lost after four simulated rainfall events (*43*). Local storm sewers in urban developments are the primary conduits for movement of products from the urban landscape to surface water. Leaves of deciduous trees treated with systemic insecticides have been recently reported as another source of contamination. Urban leaves from treated trees that leave the landscape and move into storm sewers can be a source of insecticides for urban streams. A few studies (*44,45*) have evaluated the potential impacts on leaf decomposition rates and the aquatic insects in streams that break down leaves.

Replacement, also called a secondary pest outbreak, is a phenomenon where one pest is effectively managed then another pest outbreaks. *Resurgence* is similar except that the pest that was originally being managed outbreaks after an insecticide is applied. Both of these may be tied to the final category *Impacts on Beneficial Insects and Mites*. Because the mechanisms for replacement and resurgence may involve impacts on beneficial insects and mites, they will be discussed together using case studies and experimental research. There are three common explanations for resurgence and replacement: (1) reductions of natural enemies by insecticides, (2) favorable influences of insecticides on the physiology and behavior of the pest, and (3) microbial action. Of these, the first two have the strong support among the documented cases of replacement or resurgence. It is important to remember that foliar sprays of insecticides are applied to an area or space, and not just to the pest species. Although one pest is being targeted, the application can affect other insects or mites in the area, collectively called nontargets. Nontarget species can be natural enemies in areas sprayed to control a pest, even non-plant pests like mosquitos and other biting flies. Most of these effects are documented for increases in sap-sucking insects (scale insects or whiteflies) and mites (Table 9.2). One of the first cases in the literature was a California case study conducted across 526 ha (1300 acre) of urban and suburban areas around South Lake Tahoe (*46,47*). Mosquito sprays in this area resulted in an outbreak of pine needle scale caused by a significant mortality of three parasitoid wasps of this armored scale insect. This was similar to an outbreak of whiteflies on live oak trees in coastal Mississippi following Hurricane Katrina in late August 2006. Following the hurricane, *Tetraleurodes perileuca* populations ranged from 5.7 to 35 whiteflies per leaf, and 18%–77% of the

TABLE 9.2 Field case studies documenting pest outbreaks following sprays for mosquitos, biting flies, or invasive pests.

Primary pest target	Insecticide	Outcome
Mosquito (46,47)	Malathion	Outbreak of pine needle scale, an armored scale Significant mortality of three parasitoid wasps of pine needle scale Pine foliage collected up to 15 m into the treeline from the road was lethal, 76% mortality of parasitoids, for 5 days after treatment Once the mosquito program stopped, scale populations collapsed
Biting flies (48)	DDT	Outbreaks of lecanium scale An IPM program for nuisance flies was initiated which reduced insecticide applications for fly control An immediate decline in lecanium scale coincident with the insecticide reductions
Japanese beetle (49)	DDT, carbaryl, chlordan, diazinon, dicofol	Outbreaks of citrus red mites and woolly whiteflies 5% parasitism of woolly whitefly in treated zone compared to 42.8% in nontreated zones
Mosquito (unpublished data)	Naled	Outbreak of Tetraleurodes perileuca, a whitefly Posthurricane, aerial applications of insecticide were made for mosquitos A reduction ($\geq$90%) in whitefly populations due to an increase of a parasitoid wasp

sampled foliage was infested. The next year, however, less than 10% of the leaves were infested and maximum densities of whiteflies were $\leq$10 per leaf. This was coupled with a population increase of a parasitoid wasp, which were largely absent in the sample in the preceding year (*unpublished data*).

With the exception of the most recent case, the insecticides reported in these case studies (Table 9.2) are commonly used. However, there are papers to further support the impact of urban mosquito sprays on nontarget and beneficial species. Toxicology, the science of poisons, is particularly concerned with acute or chronic exposures. Acute exposures are analogous to a stinging wasp dropping dead following a blast from an aerosol insecticide. There is usually a high dose of a toxin delivered in a short amount of time. Negative effects due to the persistence of the treatment over time following the initial application would be considered chronic. Ecotoxicology is the application of the principles of toxicology to the ecological relationships between organisms.

There are a number of studies from mosquito research that use experimental manipulation and not just observations to understand how beneficial or nontarget organisms are affected the application of urban mosquito sprays. These studies document the effects of acute exposures to mosquito applications on lady beetles (50), various small-bodied nontarget insects (51), and butterfly adults and larvae (52,53). As expected, the acute toxic effects are mainly downwind and can extend for 100 m or more from the spray vehicle (50,53). A few studies have used caterpillars as model insects to understand chronic effects. Caterpillars on host plants are commonly placed in the spray zone then kept on the treated plant until they die or complete development. These studies, so far, suggest limited chronic effects from mosquito sprays. For example, larvae of Miami blue, an ecologically threatened butterfly, were able to successfully develop into adults when fed leaves from plants fogged as part of a mosquito treatment (52). Numbers of insects in posttreatment sweep samples and those captured on sticky cards also show a recovery over time (54).

Ecotoxicological studies in urban landscape research commonly consider the impacts of routine insecticide sprays on nontarget arthropods in urban landscapes. In turfgrass research, the abundance and diversity of beneficial insects found in plots treated with insecticide are compared to those that receive no treatment on the same site (55–61). This is done for weeks, or even months, after application to determine if natural enemies impacted by treatment can recover to levels in nontreated plots. Ants, predatory beetles, spiders, earthworms, and soil-dwelling mites are common groups monitored in these studies. Other studies compare parasitism or predation (ecosystem services) in treated and nontreated plots or evaluate the acute toxicity of insecticides in laboratory or greenhouse studies (60–64). Small scale laboratory and greenhouse experiments can focus on the impacts of insecticides on specific beneficial insects and their behaviors. Those approaches are used to observe events that would not otherwise be observable under field conditions. For example, experiments that determine how certain insecticides interact with biological control agents like entomopathogenic nematodes (62,65). In addition to these published studies, research with the same insecticides in other contexts or review articles (66) commonly supplements our knowledge about ecotoxicology in urban landscapes.

The impacts of insecticide applications on natural enemies associated with ornamental plants depend on selectivity and timing of application. In urban landscapes in North America, some maintenance firms market spray programs (cover sprays) that schedule treatment, usually monthly, of many or all plants in the landscape whether pests are present or not (6,67–69). Although this is a convenient business model, urban landscapes managed with cover sprays have predictably negative consequences on beneficial insects, and are more likely to have outbreaks of scale insects (69). A recent study (21) has suggested that the negative impacts on scale insects in particular are dependent on the selectivity of the insecticides being used and timing. The United States EPA designates some insecticides as reduced risk if they have reduced mammalian toxicity, reduced hazard to beneficials, and reduced potential for environmental contamination (70). This information can also be summarized based on the severity and duration of the impact for each active ingredient (71). There may be two active ingredients that have similar efficacy against the pest but one may be more selective. Selective insecticides with a shorter residual are often called "soft" insecticides because they can be used with minimal risks to the endemic natural enemies (72). Timing of insecticide applications coincident with crawler hatch is standard dogma in urban landscape entomology. New data (21) suggest that this may, in fact, be inconsistent with conservation biological control. This study plotted the yearly appearance of life stages of three scale species in Indiana landscapes along with the abundance of their natural enemies. Insecticides timed with crawler hatch of calico scale, for example, reduced the peak abundance of parasitoids. That negative impact would be avoided if the insecticides were applied about a month earlier. For many of our more problematic pests, the occurrence or abundance of natural enemies are known. Future studies can determine the applicability of this approach to conservation biology control for other scale insects, and whether this approach is applicable in warmer climates where natural enemies may have broader periods of activity.

Up to now, we have only considered the lethal effects of insecticide applications. However, many insecticides have either indirect or sublethal effects (hormesis) that sometimes are missed when only mortality is measured (73). Sublethal effects on insects range widely but mostly alter behaviors or the physiology of insects and mites exposed to concentrations of insecticides are not lethal. Based on what was previously discussed about persistence of insecticides in plants, soil, and water, it is very likely that insects and mites would be exposed to less than lethal concentrations as the insecticide residues age. Exposures can come through soil or plants treated with insecticides. In soil, white grubs exposed to sublethal concentrations of imidacloprid or *Bacillus thuringiensis* subspecies are more susceptible to entomopathogenic nematodes by altering the normal defensive behaviors (74,75). Residues of insecticides in or on plants that are not lethal can alter behaviors which may appear like indecisiveness in herbivores and natural enemies. Green peach aphids, for example, exposed to plants sprayed with sublethal concentrations of chlorantraniliprole or imidacloprid require more search time than aphids on nontreated plants to find a suitable location to feed on the plant (76). In the same experiment, aphids on treated plants produce less honeydew indicating they may spend less time feeding in phloem and may increase xylem feeding. Adult *Tiphia vernalis*, a parasitoid of white grubs, is not killed when exposed to imidacloprid applied for grub control. However, exposed wasps have difficulty following the scent trails of grubs in soil (61). Both of these sublethal effects are effects on individual insects but they eventually translate into differences in insect populations or parasitism. As a result, the number of Japanese beetle grubs parasitized in the field in plots treated with a labeled rates or half rate of imidacloprid is significantly reduced (61).

Insecticides may also positively influence the physiology of the pest. Older publications often use the term "flare" to refer to increases in populations of insects and mites that follow applications of insecticides to plants. These effects are noted for insecticides in the carbamate, pyrethroid, organophosphate, neonicotinoid, and chlorinated hydrocarbon chemical classes (73). In research papers, these effects are noted as hormoligosis, hormesis, insecticide-induced hormoligosis, or insecticide-induced hormesis. Hormesis, a toxicological term for humans, diseases, and insects, is the most accurate term to define the reported insecticide-induced insect and mite outbreaks (73). These effects are generally defined as an insect or mite outbreak not caused by an increase in natural enemies. In some instances, populations of nontarget predators and parasitoid may also increase when exposed to insecticides or insecticide-treated plants (73,77). Some studies have shown that longevity or increased egg laying does not increase when exposed to sprays (78,79) but longevity or egg stimulation occurs if exposed to treated plants. Insecticide-mediated outbreaks of mites are documented, mainly for pest mites, on ornamental plants (79–81). However, these outbreaks do not seem to have a clear cause (dose-dependent response, increase in foliar nitrogen, etc.) for the noted effects. Two leading explanations, insecticides increase foliar quality or reduce plant defenses, are currently only anecdotally linked to these outbreaks. Imidacloprid has been shown to increase the size of foliage (82), foliar nitrogen and chlorophyll (79), but increased foliar quality was not found in cases where mite populations increased on imidacloprid-treated boxwoods or elm trees (79,81). Interestingly, some common over-the-counter formulations of the organophosphate acephate also contain the miticide fenbutatin-oxide. This may either be to increase the pest spectrum of the product or as a way to reduce "flaring" mite populations co-occurring on plants with insect pests.

The influence of microbes was previously mentioned relative to resistance. Soil microbes, however, appear to be a new frontier in plant health. Beneficial soil microbes include fungi (mycorrhizae) and bacteria that either grow inside plants (endophytes) or are applied to soil for colonization of soil in the root zone. As a result of these associations, plants experience physiological and perhaps morphological changes, except these endophytic associations create plants with no external signs or symptoms of the colonization. However, in some cases morphological changes like enhanced shoot and root growth or altered responses to abiotic (insect or disease susceptibility) or abiotic stress (drought tolerance) are possible (83). Much of the published work in this area is done with turfgrass. The application of beneficial bacteria to tall fescue or bermudagrass enables those grasses to overproduce roots in the presence of white grubs or mole crickets (84,85). The same bacteria can be mixed with insecticides commonly used in turfgrass. There is limited evidence that the bacteria and insecticide combinations encourage faster recovery from injury from root-feeding insects (85). Furthermore, when beneficial bacteria are applied with insecticides to seeds, they can enhance growth and the uptake of neonicotinoid insecticides (86). Finally, soil microbes can also masquerade as insecticide resistance through a phenomenon called microbial degradation or enhanced microbial degradation. This phenomenon is best documented with soil insecticides applied for control of insects in soil. The incidences of microbial degradation of insecticides, mostly older chemistries, in turfgrass are summarized in Ref. (35). The soil microbes on those particular sites rapidly break down those insecticides before they reach the insects. The insects in that soil context will not die following application. However, if the same insects are removed from those soils, placed in different soil, and treated with the same insecticides, they will die. Once a site is identified as having microbial degradation, the particular insecticide will no longer work when applied.

Insect conservation and the judicious use of insecticides

If you recall from the earlier chapters, insect and plant diversity are greater in suburban areas compared to rural and inner-city areas. There is a growing conservation movement among landscape designers and managers, and urban gardeners to maintain urban green spaces as refuges for state and regional insect biodiversity. The dependence on urban spaces as a haven for insect biodiversity is commonly discussed in academic circles (87,88). These same urban areas, however, overlap in the United States with an extensive presence of insecticides. Recently, two species, the western honey bee and the Monarch butterfly, have dominated much of the published literature on insect conservation in urban landscapes, particularly relative to insecticide use (Box 9.2). The concern and research efforts are related to significant population declines in overwintering Monarch butterflies in Mexico (89) and extensive annual overwintering losses in commercial western honey bee populations. In 2015, the United States EPA (90) released a preliminary report documenting 17 reported bee kill incidents associated with the use of neonicotinoid insecticides. Of those, nine were applications made to either landscape ornamental or turfgrass or in residential/urban areas. As discussed earlier, systemic insecticides move through (translocate) plant tissues. Certain neonicotinoid insecticides readily translocated into nectar and pollen in flowers, or guttation water of treated grasses and ornamental plants (29,30,91—93). Bees can also have seasonal effects on colony health from limited exposures to plant treated with certain neonicotinoid insecticides (94). In turfgrass, the primary concern is the treatment of flowering weeds like clover and dandelions (Fig. 9.2) that are common plant associates and pollinator resources in urban lawns (95). The applications of many different types of insecticides, and not just neonicotinoids, can negatively impact beneficial insects by visiting flowers treated with insecticides or flowers on treated plants (94,96). For this reason, flower

FIGURE 9.2 Bees are common inhabitants in weedy lawns. Native bees and honey bees are a common subject for conservation studies concerning insecticide use in turfgrass.

removal, via mowing, has been a recommended practice as a way to reduce hazards to beneficial insects visiting flowering weeds in lawns (*91,96,97*). In flowering ornamental plants, most insecticides now restrict applications of insecticides while plants are in flower. This has prompted research to investigate the timing of systemic insecticide applications relative to concentrations detected in the flowers. Ornamental plants have relatively high labeled rates for systemic insecticides per plant, especially when applied as a soil drench to trees and shrubs (*29,30,93*). There are few studies gauging insecticide translocation into pollen and nectar for ornamental plants. From those few studies, it is clear that treatment of plants weeks or months before bloom can translocate into pollen and nectar. Levels of insecticides detected in pollen or nectar, however, may not always exceed the current thresholds for acute or chronic toxicity to beneficial insects that visit ornamental plants (Fig. 9.3). Furthermore, the active ingredients within the neonicotinoid chemical class have different ecotoxicological profiles. Finally, many insecticide labels have a range of rates that can be legally applied. For example, applications of thiamethoxam (a neonicotinoid insecticide) at the low range of the labeled rate as a foliar spray or drench never exceed threshold levels in pollen of annual sunflower. But, drench applications of the same active ingredient to annual sunflower at any time 10 weeks or less before bloom will produce above threshold levels in pollen (*93*).

BOX 9.2 Protecting pollinators

Nectar and pollen are the rewards provided by plants in exchange for pollination. When visiting flowers, one or both are collected by worker bees and taken back to the colony. Bees are unique among insects because the adult and immature food sources come from pollen and nectar. Butterflies, true flies, and wasps use flowers but the larvae of those insects do not depend on pollen and nectar to develop.

FIGURE 9.3   Like bees, butterflies are common in urban landscapes and a common subject for conservation studies concerning insecticide use in ornamental plants.

It has been suggested that because woody plant have greater biomass, it may reduce the likelihood of exceeding threshold levels in flowers when trees and shrubs are treated with systemic insecticides (*93*). Furthermore, soil properties can impact movement of active ingredients from soil into the plant. Therefore, woody plants established in landscapes may not produce the toxic levels in the nectar or pollen seen in studies with herbaceous ornamentals potted in soilless media (*93*). There are far fewer studies with insecticide movement in woody plants established in landscapes. My laboratory has an on-going study in crape myrtle modeled after one of the few published with woody plants (*30*). Mach and Potter (*30*) treated large, multistemmed shrub hollies (*Ilex attenuata*) and summersweet shrubs (*Clethra alnifolia*) established in landscapes with two neonicotinoid insecticides—imidacloprid and dinotefuran. Applications of those two insecticides made just before bloom or the previous fall, still exceeded threshold levels in nectar for both woody plants. Applications of imidacloprid made postbloom do not produce nectar above thresholds, but that is not so for dinotefuran. With so much woody plant diversity in landscapes, it is difficult to extrapolate these results too widely. Until there are more studies, these data suggest translocation of insecticide into nectar even when applied according to the label. Applications made months or even a year prior may reduce risk but that assumes those applications are still effective against the targeted pest. These treatments may also negatively affect natural enemies also feeding on nectar or pollen in flowers on treated herbaceous plants (*29,98*), but there are no data like these in treated woody plants. These few studies assume that bees are actively foraging on these plants when in bloom. Not all flowers of woody or herbaceous plants are preferred or even commonly used by pollinators (98–100). Crape myrtle, the tree species in our research study, has a summer bloom time when few other blooms are available. The flowers do not produce nectar but are visited extensively by honey bees and bumble bees collecting pollen (100). For species like crape myrtle that have a unique or isolated bloom timing, the impact of treatments on bees may be greater than on treated, less visited flowering plants. These studies are commonly used to support special labeling for plants treated with systemic insecticide or even prohibitions on treatments with systemic insecticides, neonicotinoids in particular, by some plant retailers. However, the general public does not understand the issue or neonicotinoids. In surveys, a majority of people (56.6%) do not recognize the word "neonicotinoid" (101,102). Bee-friendly and bee-related labeling is recognized and can impact marketing. Labeling plants as "neonicotinoid-free," however, can be considered negative in the same surveys (102). Clearly more education efforts are needed to minimize knee-jerk reactions to otherwise complex interactions between bees, landscape plants, and pest management.

Highlights

- Insecticides are the most common type of pesticides in urban homes.
- Insecticide chemistries have grown increasingly effective and less hazard to humans over the last 40 years.
- Insecticide use, particularly systemic insecticide, in urban landscapes can have direct and indirect consequences for beneficial insects.
- Insecticide resistance, pest replacement and resurgence, and environmental contamination are the most common issues related to stewardship of control products and of the environment.

References

1. Grube, A.; Donaldson, D.; Kiely, T.; Wu, L. *Pesticide Industry Sales and Usage: 2006 and 2007 Market Estimates;* United States EPA, 2011; p 41.
2. Guha, N.; Ward, M. H.; Gunier, R.; Colt, J. S.; Lea, C. S.; Buffler, P. A.; Metayer, C. Characterization of Residential Pesticide Use and Chemical Formulations through Self-Report and Household Inventory: the Northern California Childhood Leukemia Study. *Environmental Health Perspectives* **2012,** *121* (2), 276–282.
3. Ryan, C.; Lewis, J. M. *Computer and Internet Use in the United States: 2015* In: *United States Census Bureau American Community Survey Reports ACS-37,* 2017; p 10.
4. Southeastern US Pest control Guide for Nursery Crops and Landscape Plantings. https://content.ces.ncsu.edu/southeastern-us-pest-control-guide-for-nursery-crops-and-landscape-plantings.
5. Insecticide Resistance Action Committee (IRAC): Mode of action classification. https://www.irac-online.org/modes-of-action/.

6. Held, D. W.; Potter, D. A. Prospects for Managing Turfgrass Pests with Reduced Chemical Inputs. *Annual Review of Entomology* **2012,** *57,* 329−354.

7. Nielsen, D. G. Studying Biology and Control of Borers Attacking Woody Plants. *Bulletin of the Entomological Society of America* **1981,** *27* (4), 251−260.

8. Quesada, C. R.; Sadof, C. S. Efficacy of Horticultural Oil and Insecticidal Soap against Selected Armored and Soft Scales. *HortTechnology* **2017,** *27* (5), 618−624.

9. Buchholz, A. Characterization of the Diffusion of Non-electrolytes across Plant Cuticles: Properties of the Lipophilic Pathway. *Journal of Experimental Botany* **2006,** *57* (11), 2501−2513.

10. Buchholz, A.; Trapp, S. How Active Ingredient Localisation in Plant Tissues Determines the Targeted Pest Spectrum of Different Chemistries. *Pest Management Science* **2016,** *72,* 929−939.

11. Smitley, D. R.; Herms, D. A.; Davis, T. W. Efficacy of Soil-Applied Neonicotinoid Insecticides for Long-Term Protection against Emerald Ash Borer (Coleoptera: Buprestidae). *Journal of Economic Entomology* **2015,** *108* (5), 2344−2353.

12. Held, D. W.; Parker, S. A. Efficacy of Soil-Applied Neonicotinoid Insecticides against Azalea Lace Bug, *Stephanitis pyroides,* in the Landscape. *Florida Entomologist* **2011,** *94* (3), 599−607.

13. Szczepaniec, A.; Creary, S. F.; Laskowski, K. L.; Nyrop, J. P.; Raupp, M. J. Neonicotinoid Insecticide Imidacloprid Causes Outbreaks of Spider Mites on Elm Trees in Urban Landscapes. *PLoSOne* **2011,** *6* (5), e20018.

14. Eisenback, B. M.; Salom, S. M.; Kok, L. T.; Lagalante, A. F. Impacts of Trunk and Soil Injections of Low Rates of Imidacloprid on Hemlock Woolly Adelgid (Hemiptera: Adelgidae) and Eastern Hemlock (Pinales: Pinaceae) Health. *Journal of Economic Entomology* **2014,** *107* (1), 250−258.

15. Williamson, R. C.; Liesch, P. J.; Obear, G. Residual Activity of Chlorantraniliprole and Other Turfgrass Insecticides against Black Cutworm (Lepidoptera: Noctuidae). *International Turfgrass Society Research Journal* **2013,** *12,* 1−5.

16. Tomizawa, M.; Casida, J. E. Neonicotinoid Insecticide Toxicology: Mechanisms of Selective Action. *Annual Review of Pharmacology and Toxicology* **2005,** *45* (1), 247−268.

17. Casida, J. E. Neonicotinoids and Other Insect Nicotinic Receptor Modulators: Progress and Prospects. *Annual Review of Entomology* **2018,** *63,* 125−144.

18. Barry, J. D.; Portillo, H. E.; Annan, I. B.; Cameron, R. A.; Clagg, D. G.; Dietrich, R. F.; Watson, L. J.; Leighty, R. M.; Ryan, D. L.; McMillan, J. A.; Swain, R. S.; Kaczmarczyk, R. A. Movement of Cyantraniliprole in Plants after Foliar Applications and its Impact on the Control of Sucking and Chewing Insects. *Pest Management Science* **2015,** *71,* 395−403.

19. Tofangsazi, N.; Cherry, R. H.; Beeson, R. C.; Arthurs, S. P. Concentration−response and Residual Activity of Insecticides to Control *Herpetogramma phaeopteralis* (Lepidoptera: Crambidae) in St. Augustinegrass. *Journal of Economic Entomology* **2015,** *108* (2), 730−735.

20. Bonmatin, J. M.; Giorio, C.; Girolami, V.; Goulson, D.; Kreutzweiser, D. P.; Krupke, C.; Marzaro, M.; Mitchell, E. A. D.; Noome, D. A.; Simon-Delso, N.; Tapparo, A. Environmental Fate and Exposure; Neonicotinoids and Fipronil. *Environmental Science and Pollution Research* **2015,** *22,* 35−67.

21. Quesada, C. R.; Sadof, C. S. Field Evaluation of Insecticide and Application Timing on Natural Enemies of Selected Armored and Soft Scales. *Biological Control* **2019,** *133,* 81−90.

22. Doccola, J. J.; Wild, P. M. Tree Injection as an Alternative Method of Insecticide Application. In *Insecticides-basic and other applications;* Soloneski, S., Larramendy, M., Eds.; InTech: Rijeka, Croatia, 2012; pp 61−78.

23. Doccola, J. J.; Smitley, D. R.; Davis, T. W.; Aiken, J. J.; Wild, P. M. Tree Wound Responses Following Systemic Insecticide Trunk Injection Treatments in Green Ash (*Fraxinus pennsylvanica* Marsh.) as Determined by Destructive Autopsy. *Arboriculture & Urban Forestry* **2011,** *37* (1), 6−12.

24. Tattar, T.; Dotson, J. A.; Ruizzo, M. S.; Steward, V. B. Translocation of Imidacloprid in Three Tree Species when Trunk- and Soil-Injected. *Journal of Arboriculture* **1998,** *24* (1), 54−56.

25. Benton, E.; Grant, J. F.; Cowles, R.; Webster, J.; Nichols, R.; Lagalante, A.; Coots, C. Assessing Relationships between Tree Diameter and Long-Term Persistence of Imidacloprid and Olefin to Optimize Imidacloprid Treatments on Eastern Hemlock. *Forest Ecology and Management* **2016,** *370,* 12−21.

26. Ugine, T. A.; Gardescu, S.; Hajek, A. E. The Within-Season and Between-Tree Distribution of Imidacloprid Trunk-Injected into *Acer platanoides* (Sapindales: Sapindaceae). *Journal of Economic Entomology* **2013,** *106* (2), 874−882.

27. Mota-Sanchez, D.; Cregg, B. M.; McCullough, D. G.; Poland, T. M.; Hollingworth, R. M. Distribution of Trunk-Injected 14C-Imidacloprid in Ash Trees and Effects on Emerald Ash Borer (Coleoptera: Buprestidae) Adults. *Crop Protection* **2009,** *28* (8), 665-661.

28. Nix, K.; Lambdin, P.; Grant, J.; Coots, C.; Merten, P. Concentration Levels of Imidacloprid and Dinotefuran in Five Tissue Types of Black Walnut. *Juglans nigra. Forests* **2013,** *4,* 887−897.

29. Krischik, V.; Rogers, M.; Gupta, G.; Varshney, A. Soil-applied Imidacloprid Translocates to Ornamental Flowers and Reduces Survival of Adult *Coleomegilla maculata, Harmonia axyridis,* and *Hippodamia convergens* Lady Beetles, and Larval *Danaus plexippus* and *Vanessa cardui* Butterflies. *PLoS One* **2015,** *10* (3), e0119133.

30. Mach, B. M.; Bondarenko, S.; Potter, D. A. Uptake and Dissipation of Neonicotinoid Residues in Nectar and Foliage of Systemically Treated Woody Landscape Plants. *Environmental Toxicology & Chemistry* **2018,** *37,* 860−870.

31. Chandra, A.; Reinert, J. A.; LaMantia, J.; Pond, J. B.; Huff, D. R. Genetic Variability in Populations of the Southern Chinch Bug, *Blissus insularis,* Assessed Using AFLP Analysis. *Journal of Insect Science* **2011,** *11,* 173.

32. McKenzie, C.; Osborne, L. *Bemisia tabaci* MED (Q Biotype) (Hemiptera: Aleyrodidae) in Florida Is on the Move to Residential Landscapes and May Impact Open-Field Agriculture. *Florida Entomologist* **2017,** *100,* 481−484.

33. Dickey, A. M.; Kumar, V.; Hoddle, M. S.; Funderburk, J. E.; Morgan, J. K.; Jara-Cavieres, A.; Shatters, R. G., Jr.; Osborne, L. S.; McKenzie, C. L. The *Scirtothrips dorsalis* Species Complex: Endemism and Invasion in a Global Pest. *PLoS One* **2015,** *10* (4), e0123747.

34. Borel, B. CRISPR, Microbes and More Are Joining the War against Crop Killers. *Nature* **2017,** *543,* 302−304.

35. Gelernter, W. Insect Pests of Turfgrass: Management Challenges in a Changing Environment. In *Handbook of Turfgrass Insects;* Brandenburg, R. L., Freeman, C. P., Eds., 2nd ed.; Entomological Society of America: Maryland, 2012; pp 2−8.

36. Adesanya, A.; Liu, N.; Held, D. W. Multiple Detoxification Enzymes Facilitate Generalism in Adult Japanese Beetles, *Popillia japonica* Newman. *Journal of Insect Physiology* **2016,** *88,* 55−62.

37. Xu, Y.; Buss, E. A.; Boucias, D. G. Environmental Transmission of the Gut Symbiont *Burkholderia* to Phloem-Feeding *Blissus insularis*. *PLoS One* **2016**, *11* (8), e0161699.

38. Xu, Y.; Buss, E. A.; Boucias, D. G. Impacts of Antibiotic and Bacteriophage Treatments on the Gut-Symbiont-Associated *Blissus insularis* (Hemiptera: Blissidae). *Insects* **2016**, *7*, 61.

39. Miersma, N. A.; Pepper, C. A.; Anderson, T. A. Organochlorine Pesticides in Elementary School Yards along the Texas-Mexico Border. *Environmental Pollution* **2003**, *126*, 65−71.

40. Riederer, A. M.; Smith, K. D.; Barr, D. B.; Hayden, S. W.; Hunter, R. E., Jr.; Ryan, P. B. Current and Historically Used Pesticides in Residential Soil from 11 Homes in Atlanta, Georgia, USA. *Archives of Environmental Contamination and Toxicology* **2010**, *58*, 908−917.

41. López, R.; Held, D. W.; Potter, D. A. Management of a Mound-Building Ant, *Lasius neoniger* Emery, on Golf Course Putting Greens Using Delayed Action Baits or Fipronil. *Crop Science* **2000**, *40*, 511−517.

42. Coles, J. F.; McMahon, G.; Bell, A. H.; Brown, L. R.; Fitzpatrick, F. A.; Scudder Eikenberry, B. C.; Woodside, M. D.; Cuffney, T. F.; Bryant, W. L.; Cappiella, K.; Fraley-McNeal, L.; Stack, W. P. Effects of Urban Development on Stream Ecosystems in Nine Metropolitan Study Areas across the United States: U.S. *Geological Survey Circular* **2012**, *1373*, 138.

43. Armbrust, K. L.; Peeler, H. B. Effects of Formulation on the Run-off of Imidacloprid from Turf. *Pest Management Science* **2002**, *58*, 702−706.

44. Kreutzweiser, D. P.; Good, K. P.; Chartrand, D. T.; Scarr, T. A.; Thompson, D. G. Are Leaves that Fall from Imidacloprid-Treated Maple Trees to Control Asian Longhorned Beetles Toxic to Non-target Decomposer Organisms? *Journal of Environmental Quality* **2008**, *37* (2), 639−646.

45. Englert, D.; Zubrod, J. P.; Pietz, S.; Stefani, S.; Krauss, M.; Schulz, R.; Bundschuh, M. Relative Importance of Dietary Uptake and Waterborne Exposure for a Leaf-Shredding Amphipod Exposed to Thiacloprid-Contaminated Leaves. *Scientific Reports* **2017**, *7*, 16182.

46. Luck, R. F.; Dahlstein, D. L. Natural Decline of a Pine Needle Scale (*Chionaspis pinifoliae* [Fitch]), Outbreak at South Lake Tahoe, California Following Cessation of Adult Mosquito Control with Malathion. *Ecology* **1975**, *56*, 893−904.

47. Dahlsten, D. L.; Garcia, R.; Prine, J. E.; Hunt, R. Insect Problems in Forest Recreation Areas: Pine Needle scale...Mosquitoes. *California Agriculture* **1969**, *31*, 4−6.

48. Merritt, R. W.; Kennedy, M. K.; Gersabeck, E. F. Integrated Pest Management of Nuisance and Biting Flies in a Michigan Resort: Dealing with Secondary Pest Outbreaks. In *Urban Entomology: Interdisciplinary Perspectives;* Frankie, G. W., Koehler, C. S., Eds.; Praeger: New York, 1983; pp 277−299.

49. DeBach, P.; Rose, M. Environmental Upsets Caused by Chemical Eradication. *California Agriculture* **1977**, *31*, 8−10.

50. Peterson, R. K. D.; Preftakes, C. J.; Bodin, J. L.; Brown, C. R.; Piccolomini, A. M.; Schleier, J. J. Determinants of Acute Mortality of *Hippodamia convergens* (Coleoptera: Coccinellidae) to Ultra-low Volume Permethrin Used for Mosquito Management. *PeerJ* **2016**, *4*, e2167.

51. Boyce, W. M.; Lawler, S. P.; Schultz, J. M.; McCauley, S. J.; Kimsey, L. S.; Niemela, M. K.; Nielsen, C. F.; Reisen, W. K. Nontarget Effects of the Mosquito Adulticide Pyrethrin Applied Aerially during a West Nile Virus Outbreak in an Urban California Environment. *Journal of the American Mosquito Control Association* **2007**, *23* (3), 335−339.

52. Zhong, H.; Hribar, L. J.; Daniels, J. C.; Feken, M. A.; Brock, C.; Trager, M. D. Aerial Ultra-low-volume Application of Naled: Impact on Nontarget Imperiled Butterfly Larvae

(*Cyclargus thomasi* Bethunebakeri) and Efficacy against Adult Mosquitoes (*Aedes Taeniorhynchus*). *Environmental Entomology* **2010**, *39*, 1961−1972.

53. Oberhauser, K. S.; Manweiler, S. A.; Lelich, R.; Blank, M.; Batalden, R. V.; De Anda, A. Impacts of Ultra-low Volume Resemethrin Applications on Non-target Insects. *Journal of the American Mosquito Control Association* **2009**, *25*, 83−93.

54. Davis, R. S.; Peterson, R. K. D. Effects of Single and Multiple Applications of Mosquito Insecticides on Non-target Arthropods. *Journal of the American Mosquito Control Association* **2008**, *24*, 270−280.

55. Cockfield, S. D.; Potter, D. A. Predation of Sod Webworm (Lepidoptera: Pyralidae) Eggs as Affected by Chlorpyrifos Application to Kentucky Bluegrass Turf. *Journal of Economic Entomology* **1984**, *77*, 1542−1544.

56. Braman, S. K.; Pendley, A. F. Relative and Seasonal Abundance of Beneficial Arthropods in Centipede Grass as Influenced by Management Practices. *Journal of Economic Entomology* **1993**, *86*, 494−504.

57. Kunkel, B. A.; Held, D. W.; Potter, D. A. Impact of Halofenozide, Imidacloprid, and Bendiocarb on Beneficial Invertebrates and Predatory Activity in Turfgrass. *Journal of Economic Entomology* **1999**, *92* (4), 922−930.

58. Peck, D. C. Long-term Effects of Imidacloprid on the Abundance of Surface- and Soil-active Nontarget Fauna in Turf. *Agricultural and Forest Entomology* **2009**, *11*, 405−419.

59. Larson, J. L.; Redmond, C. T.; Potter, D. A. Comparative Impact of an Anthranilic Diamide and Other Insecticidal Chemistries on Beneficial Invertebrates and Ecosystem Services in Turfgrass. *Pest Management Science* **2012**, *68*, 740−748.

60. Kunkel, B. A.; Held, D. W.; Potter, D. A. Lethal and Sublethal Effects of Bendiocarb, Halofenozide, and Imidacloprid on *Harpalus pennsylvanicus* (Coleoptera:Carabidae) Following Different Modes of Exposure in Turfgrass. *Journal of Economic Entomology* **1999**, *94*, 60−67.

61. Rogers, M. E.; Potter, D. A. Effects of Spring Imidacloprid Application for White Grub Control on Parasitism of Japanese Beetle (Coleoptera: Scarabaeidae) by Tiphia Vernalis (Hymenoptera: Tiphiidae). *Journal of Economic Entomology* **2003**, *96*, 1412−1419.

62. Koppenhöfer, A. M.; Fuzy, E. M. Effect of the Anthranilic Diamide Insecticide, Chlorantraniliprole, on *Heterorhabditis bacteriophora* (Rhabditida: Heterorhabditidae) Efficacy against White Grubs (Coleoptera: Scarabaeidae). *Biological Control* **2008**, *45*, 93−102.

63. Larson, J. L.; Redmond, C. T.; Potter, D. A. Impacts of Neonicotinoid, Neonicotinoid-Pyrethroid Premix, and Anthranilic Diamide Insecticides on Four Species of Turf-Inhabiting Beneficial Insects. *Ecotoxicology* **2014**, *23*, 252−259.

64. Oliver, J. B.; Reding, M. E.; Moyseenlko, J. J.; Klein, M. G.; Mannion, C. M.; Bishop, B. Survival of Adult *Tiphia vernalis* (Hymenoptera: Tiphiidae) after Insecticide, Fungicide, and Herbicide Exposure in Laboratory Bioassays. *Journal of Economic Entomology* **2006**, *99* (2), 288−294.

65. Barbara, K. A.; Buss, E. A. Integration of Insect Parasitic Nematodes (Rhabditida: Steinernematidae) with Insecticides for Control of Pest Mole Crickets (Orthoptera: Gryllotalpidae: Scapteriscus spp.). *Journal of Economic Entomology* **2005**, *98*, 689−693.

66. Pisa, L. W.; Amaral-Rogers, V.; Belzunces, L. P.; Bonmatin, J. M.; Downs, C. A.; Goulson, D.; Kreutzweiser, D. P.; Krupke, C.; Liess, M.; McField, M.; Morrissey, C. A.; Noome, D. A.; Settele, J.; Simon-Delso, N.; Stark, J. D.; Van der Sluijs, J. P.; Van Dyck, H.; Wiemers, M. Effects of Neonicotinoids and Fipronil on Non-target Invertebrates. *Environmental Science and Pollution Research* **2015**, *22*, 68−102.

67. Braman, S. K.; Latimer, J. G.; Robacker, C. D. Factors Influencing Pesticide Use and Integrated Pest Management Implementation in Urban Landscapes: a Case Study in Atlanta. *HortTechnology* **1998**, *8*, 145–149.

68. Stewart, C. D.; Braman, S. K.; Sparks, B. L.; Williams-Woodward, J. L.; Wade, G. L.; Latimer, J. G. Comparing an IPM Pilot Program to a Traditional Cover Spray Program in Commercial Landscapes. *Journal of Economic Entomology* **2002**, *95* (4), 789–796.

69. Raupp, M. J.; Holmes, J. J.; Sadof, C.; Shrewsbury, P.; Davidson, J. A. Effects of Cover Sprays and Residual Pesticides on Scale Insects and Natural Enemies in Urban Forests. *Journal of Arboriculture* **2001,** *29* (4), 203–214.

70. US EPA, Reduced Risk and Organophosphate Alternative Decisions for Conventional Pesticides https://www.epa.gov/pesticide-registration/reduced-risk-and-organophosphate-alternative-decisions-conventional.

71. UC Pest Management Guidelines: Floriculture and Ornamental Nurseries: Biological Control http://ipm.ucanr.edu/PMG/r280390111.html#TABLE1.

72. Frank, S. D. Reduced Risk Insecticides to Control Scale Insects and Protect Natural Enemies in the Production and Maintenance of Urban Landscape Plants. *Environmental Entomology* **2012,** *41* (2), 377–386.

73. Guedes, R. N.; Cutler, G. C. Insecticide-induced Hormesis and Arthropod Pest Management. *Pest Management Science* **2014,** *70*, 690–697.

74. Koppenhöfer, A. M.; Wilson, M.; Brown, I.; Kaya, H.; Gaugler, R. Biological Control Agents for White Grubs (Coleoptera: Scarabaeidae) in Anticipation of the Establishment of the Japanese Beetle in California. *Journal of Economic Entomology* **2000,** *93*, 71–80.

75. Koppenhöfer, A. M.; Grewal, P. S.; Kaya, H. K. Synergism of Imidacloprid and Entomopathogenic Nematodes Against White Grubs: The Mechanism. *Entomologia Experimentalis et Applicata* **2000,** *94*, 283–293.

76. Zeng, X.; He, Y.; Wu, J.; Tang, Y.; Gu, J.; Ding, W.; Zhang, Y. Sublethal Effects of Cyantraniliprole and Imidacloprid on Feeding Behavior and Life Table Parameters of *Myzus persicae* (Hemiptera: Aphididae). *Journal of Economic Entomology* **2016,** *109* (4), 1595–1602.

77. James, D. G. Imidacloprid Increases Egg Production in *Amblyseius victoriensis* (Acari: Phytoseiidae). *Experimental & Applied Acarology* **1997,** *21*, 75–82.

78. Saini, R. S.; Cutkomp, L. K. The Effects of DDT and Sublethal Doses of Dicofol on Reproduction of the Two-Spotted Spider Mite. *Journal of Economic Entomology* **1966,** *59* (2), 249–253.

79. Szczepaniec, A.; Raupp, M. J. Direct and Indirect Effects of Imidacloprid on Fecundity and Abundance of *Eurytetranychus buxi* (Acari: Tetranychidae) on Boxwoods. *Experimental & Applied Acarology* **2013,** *59*, 307–318.

80. Szczepaniec, A. *Mechanisms Underlying Outbreaks of Spider Mites Following Applications of Imidacloprid.* Ph.D. dissertation; University of Maryland, 2009.

81. Raupp, M. J.; Webb, R.; Szczepaniec, A.; Booth, D.; Ahern, R. Incidence, Abundance, and Severity of Mites on Hemlocks Following Applications of Imidacloprid. *Journal of Arboriculture* **2004,** *30*, 108–113.

82. Gupta, G.; Krischik, V. A. Professional and Consumer Insecticides for Management of Adult Japanese Beetle on Hybrid Tea Rose. *Journal of Economic Entomology* **2007,** *100* (3), 830–837.

83. Coy, R. M. *Plant Growth-Promoting Rhizobacteria (PGPR) Mediate Interactions Between Abiotic and Biotic Stresses in Cool- and Warm-Season Grasses.* PhD dissertation; Auburn University, 2017.

84. Coy, R. M.; Held, D. W.; Kloepper, J. W. Rhizobacterial Treatments of Tall Fescue and Bermudagrass Increases Tolerance to Damage from White Grubs. *Pest Management Science* **2019**. In press.

85. Coy, R. M.; Held, D. W.; Kloepper, J. W. Rhizobacterial Treatment of Bermudagrass Alters Tolerance to Damage from Tawny Mole Crickets (Neoscapteriscus *vicinus* Scudder). *Pest Management Science* **2019**. In press.

86. Myresiotis, C. K.; Vryzas, Z.; Papadopoulou-Mourkidou, E. Effect of Specific Plant-Growth-Promoting Rhizobacteria on Growth and Uptake of Neonicotinoid Insecticide Thiamethoxam in Corn (*Zea mays* L.) Seedlings. *Pest Management Science* **2015**, *71*, 1258−1266.

87. Faeth, S. H.; Bang, C.; Saari, S. Urban Biodiversity: Patterns and Mechanisms. *Annals of the New York Academy of Sciences* **2011**, *1223*, 69−81.

88. Mata, L.; Goula, M.; Hahs, A. K. Conserving Insect Assemblages in Urban Landscapes: Accounting for Species-specific Responses and Imperfect Detection. *Journal of Insect Conservation* **2014**, *18*, 885−894.

89. Brower, L. P.; Taylor, O. R.; Williams, E. H.; Slauback, D. A.; Zubieta, R. R.; Ramírez, M. I. Decline of Monarch Butterflies Overwintering in Mexico: Is the Migratory Phenomenon at Risk? *Insect Conservation and Diversity* **2012**, *5*, 95−100.

90. US EPA Preliminary Pollinator Assessment to Support the Registration Review of Imidacloprid. https://www.regulations.gov/document?D=EPA-HQ-OPP-2008-0844-0140.

91. Larson, J. L.; Redmond, C. T.; Potter, D. A. Mowing Mitigates Bioactivity of Neonicotinoid Insecticide in Nectar of Flowering Lawn Weeds and Turfgrass Guttation. *Environmental Toxicology & Chemistry* **2015**, *34* (1), 127−132.

92. McCurdy, J.; Held, D. W.; Gunn, J. M.; Barickman, T. C. Dew from Warm- Season Turfgrass as a Possible Route for Pollinator Exposure to Lawn-Applied Imidacloprid. *Crop Forage Turfgrass Science* **2017**; https://doi.org/10.2134/cftm2016.09.0063.

93. Cowles, R. S.; Eitzer, B. D. Residues of Neonicotinoid Insecticide in Pollen and Nectar from Model Plants. *Journal of Environmental Horticulture* **2017**, *35* (1), 24−34.

94. Larson, J. L.; Redmond, C. T.; Potter, D. A. Assessing Insecticide Hazard to Bumble Bees Foraging on Flowering Weeds in Treated Lawns. *PLoS One* **2013**, *8* (6), e66375.

95. Larson, J. L.; Kesheimer, A. J.; Potter, D. A. Pollinator Assemblages on Dandelions and White Clover in Urban and Suburban Lawns. *Journal of Insect Conservation* **2014**, *18*, 863−873.

96. Gels, J. A.; Held, D. W.; Potter, D. A. Hazards of Insecticides to the Bumble Bees Bombus Impatiens (Hymenoptera: Apidae) Foraging on Flowering White Clover in Turf. *Journal of Economic Entomology* **2002**, *95*, 722−728.

97. Larson, J. L.; Dale, A.; Held, D.; McGraw, B.; Richmond, D. S.; Wickings, K.; Williamson, R. C. Optimizing Pest Management Practices to Conserve Pollinators in Turf Landscapes: Current Practices and Future Research Needs. *Journal of Integrated Pest Management* **2017**, *8* (1), 1−10.

98. Mach, B. M.; Potter, D. A. Quantifying Bee Assemblages and Attractiveness of Flowering Woody Landscape Plants for Urban Pollinator Conservation. *PLoS One* **2018**, *13* (12), e0208428.

99. Held, D. W.; Chen, Y.; Knox, G.; Pemberton, B.; Carroll, D.; Layton, B. *Protecting Pollinators in Urban Areas: Use of Flowering Plants. ANR-2419;* Alabama Cooperative Extension System, 2017; p 8.

100. Riddle, T. C.; Mizell, R. F., III Use of Crape Myrtle, Lagerstroemia (Myrtles: Lythraceae), Cultivars as a Pollen Source by Native and Non-native Bees (Hymenoptera: Apidae) in Quincy, Florida. *Florida Entomologist* **2016,** *99* (1), 38−46.
101. Wollaeger, H. M.; Getter, K. L.; Behe, B. K. Consumer Preferences for Traditional Neonicotinoid-free, Bee-Friendly, or Biological Control Pest Management Practices on Floricultural Crops. *HortScience* **2015,** *50* (5), 721−732.
102. Rihn, A.; Khachatryan, H. Does Consumer Awareness of Neonicotinoid Insecticides Influence Their Preferences for Plants? *HortScience* **2016,** *51* (4), 388−393.

Index

Printed in the United States
By Bookmasters